Meandering Methodologies, Deviant Disciplines

Four Years of City Art Collaboratory

Shanai Matteson, Editor

Meandering Methodologies, Deviant Disciplines

Four Years of City Art Collaboratory

Shanai Matteson, Editor
Colleen Sheehy, Associate Editor

PUBLIC ART
SAINT PAUL

Published by Public Art Saint Paul, Saint Paul, Minnesota, 2016
ISBN 978-0-692-76841-9
Library of Congress Control Number: 2016951084

Designed by Colin Kloecker (Works Progress Studio).
Typefaces used are Lyon Text, designed by Kai Bernau for Commercial Type, and Aperçu, designed by Colophon Foundry.
Photo on front cover by Sean Smuda. Photo on back cover by Andrea Steudel.

Copyedited by Melissa Constantine.

Generous support for this project provided by the National Endowment for the Arts - ArtWorks, District Energy, The Saint Paul Foundation, F.R. Bigelow Foundation, Mardag Foundation, The John and Ruth Huss Fund of The Saint Paul Foundation, and the voters of Minnesota, through a Minnesota State Arts Board Operating Support grant, thanks to a legislative appropriation from the arts and cultural heritage fund.

MINNESOTA
STATE ARTS BOARD

CLEAN
WATER
LAND &
LEGACY
AMENDMENT

ART WORKS.

National
Endowment
for the Arts
arts.gov

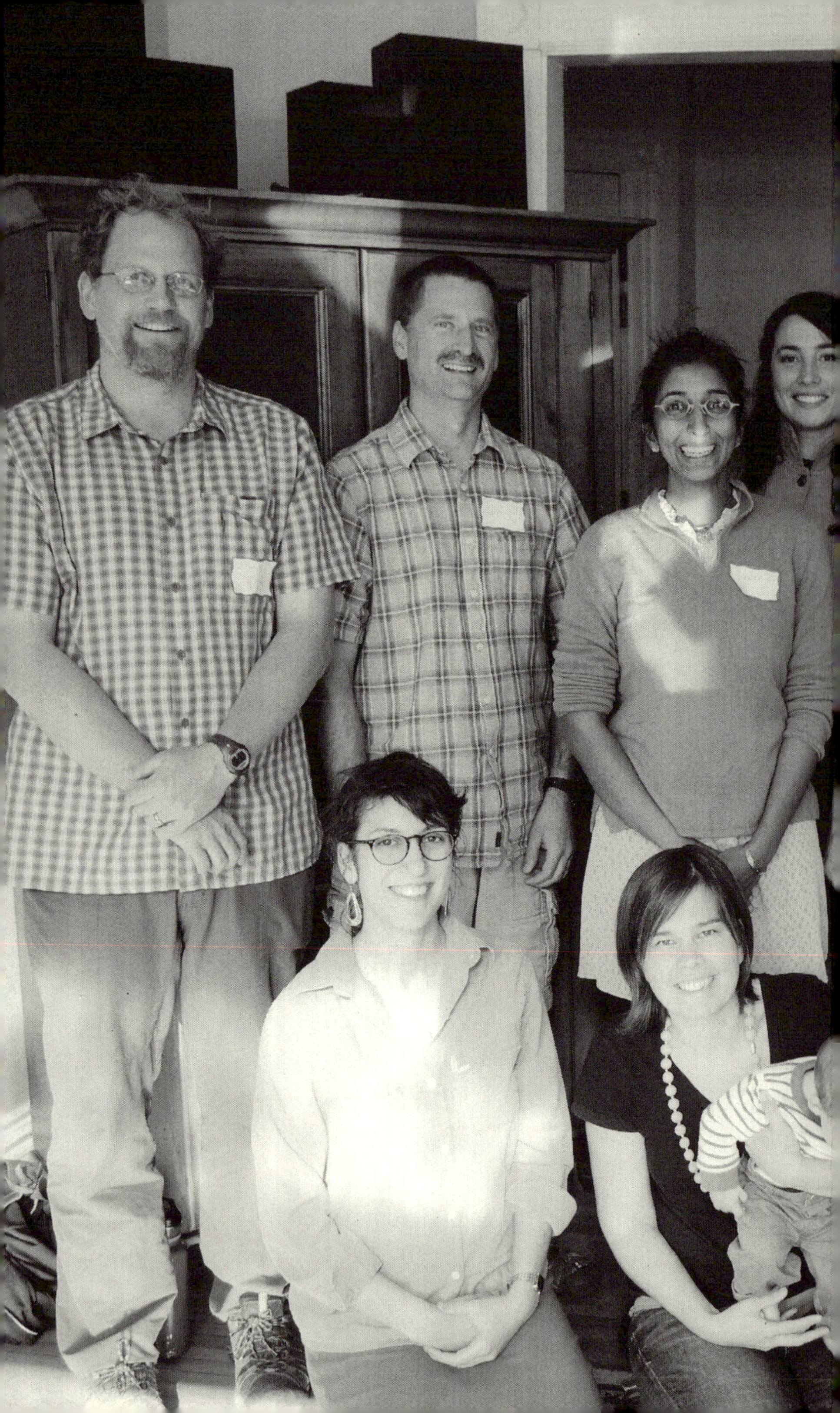

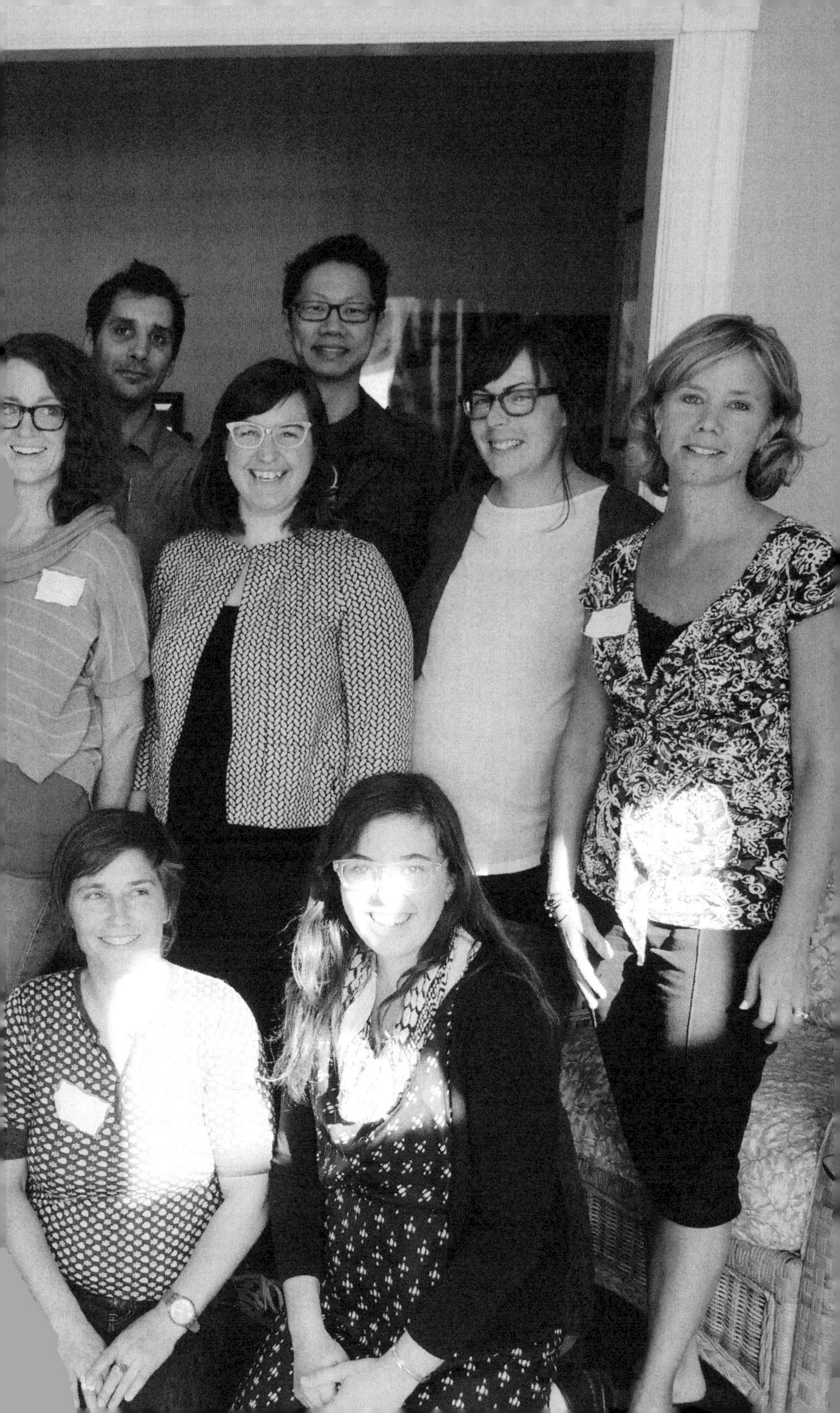

Preceding pages: The first group meeting of City Art Collaboratory Fellows at the home of then-President of Public Art Saint Paul, Christine Podas-Larson, 2013. Photo: Kent Larson.

Introduction: Incubating an Art-Science Hybrid

Colleen Sheehy

Executive Director of Public Art Saint Paul Colleen Sheehy supports artists who collaborate in the planning and design of public places, structures, and systems, and advocates for the vital role of art in civic life.

This book of essays and reflections by participating artists and scientists in the City Art Collaboratory program represents vital aspects of the core values of Public Art Saint Paul, which have evolved over the course of our nearly thirty-year history of creating public spaces, public art, and public life in the city. We believe in the power of artists to innovate and to find new ways of thinking, experiencing, and understanding the world, in turn creating richer lives for us all. We believe in cross-sector, cross-cultural, and cross-disciplinary conversations and collaborations, as we connect artists with a wide range of partners, including neighborhood leaders, small business owners, engineers, scientists from varied fields, and other professionals who help to create and care for city systems. Begun in 2012 and supported by generous grants from the National Endowment for the Arts, the City Art Collaboratory facilitated field trips to, among other places, energy and water treatment plants, city maintenance shops, and Dakota sites along the Mississippi River. Collaboratory Fellows took time to converse and get to know each other, responding to their sites of exploration and to each other's fields of thought and research. This book offers an opportunity for the Fellows to reflect on what they've learned together and what new ideas and questions have evolved from their forays and conversations—and to share these insights with a broad audience.

Christine Podas-Larson, founder and director of Public Art Saint Paul for twenty-eight years, advanced this art-science vision within the field of public art. She pioneered a unique public art program that placed artists within City departments in Saint Paul, beginning in 2005 as City Artists in Residence. The sculptor Steven Woodward was the inaugural artist, followed by Marcus Young, Amanda Lovelee, Sarah West, and Aaron Dysart. Each artist worked with engineers, landscape architects, water specialists, and other professionals in the city. Christine was also inspired by the growing awareness of human impact on the environment and knew of artists working on issues of environmental sustainability. These threads led her to propose the forerunner to the Collaboratory, the

Sustainable Artmaking Fellowships in 2008, followed by the Hunting and Gathering Walks in 2010. Her commitment to the process and perspective of artists permeated this experimental think tank and generated opportunities for artists and scientists to work together.

When I came to Public Art Saint Paul in summer 2015, taking the helm when Podas-Larson retired, I was compelled by these ongoing art-science-sustainability initiatives. I had worked in public art in a museum context, and had also forged a strong interdisciplinary methodology, informed by my graduate field of American Studies. Further, my work at the University of Minnesota's Weisman Art Museum reached across departments and disciplinary boundaries. The edges of disciplines always seemed to me to be where the most interesting work was happening, like the subatomic spaces where electrons interact to form new substances. For these reasons, the City Art Collaboratory was a significant factor in my decision to lead Public Art Saint Paul. I thank Christine Podas-Larson for her amazing vision, giving rise to a unique program that offers artists time and financial support to explore ideas, ask questions, discover places, and build new relationships. I regard the Collaboratory as an almost utopian endeavor that makes space for creative people to connect, expand, and process in an opened-ended way. This is a rarity, particularly outside of the centers or institutes found at universities. Yet it provides a model for other, much needed opportunities to advance cross-disciplinary ideas and make change. We hope readers of this book will be inspired to create their own versions of the City Art Collaboratory, wherever they may be.

This book shares the processes, conversations, insights, and impacts of our experimental program with the field of public art and with those working at the intersections of art and science. It will also be valuable to those who work for city governments or are interested in issues of urbanism, place, and sustainability. The contents here convey the character of the endeavor and the commitments of the individuals involved. Essays range widely from short reflections on experiences to theoretical think pieces composed by teams of artists-scientists. We also invited two writers outside of the Collaboratory to contribute essays. Christina Schmid is an art critic based in the Twin Cities who teaches in the Department of Art at the University of Minnesota. She spent time with Collaboratory members and wrote an insightful essay on how the group's commitment to place is very much against the grain of the contemporary art world. Critic, teacher, and artist Dan S. Wang provides a framework for interpreting experiences at Coldwater Spring, a site on the Mississippi River that has a charged history, both recent and long, between Dakota people, to whom the site is sacred, and Euro-Americans who settled there and altered the landscape. As one of the Compass Collaborators, a group of artists concerned with art, culture, and politics in this region, Wang had used a similar field trip methodology with his group to explore charged sites, and had published writing about those experiences. I express deep thanks to all of the authors, whose work enriches this book.

Shanai Matteson has served as the Director of City Art Collaboratory for four years and led this publication effort as general editor and essayist. I thank Shanai for the inspired leadership she has provided for the Collaboratory program and Fellows. Her two essays here provide an overview of the precedents of the Collaboratory at Public Art Saint Paul, going back to 2008. She also chronicles the ways in which having time and space to "meander" helped to foster a "community of practice" among the cohort of Fellows and their wider networks.

While thus far the City Art Collaboratory has been devoted to a process, not a product, some real world projects have emerged from different members; many of these are truly inventive and reveal the power of art-science collaborations. Amanda Lovelee was a member of the Collaboratory, and was subsequently hired to be a City Artist with Public Art Saint Paul in 2013. Inspired to work with a scientist from her experience with the Collaboratory, she teamed up with ecologist Adam Kay to create a major new work of temporary public art, *Urban Flower Field*, the subject of their conversation in this book. Not only has it been a popular new place for people to enjoy the outdoors, it has also hosted science experiments that Kay designed and his students have implemented. *Urban*

Flower Field has been a three-year transitional space with the possibility to extend into the future.

Other Collaboratory Art Fellows, Aaron Dysart, Emily Stover, and Asia Ward, developed *The Plume Project*, materializing an idea that emerged from a Collaboratory field trip to District Energy in downtown Saint Paul. The energy plant is one of the most innovative in the country, using wood waste to heat and cool water that is piped to buildings in downtown Saint Paul. With funds from the Knight Foundation's Art Challenge, they created a fascinating and spectacular vision by illuminating the steam plume emitted from District Energy's water-generated heating and cooling. Each artist created a unique lighting design to illuminate the steam plume, involving interactive components with the public or data visualization that generated changing colors and images.

Finally, Shanai Matteson and her partner Colin Kloecker—together the public art and design studio Works Progress—launched *Water Bar* to bring attention to issues of water stewardship. After introducing this social/art/science engagement project at festivals and events through temporary, portable water bars, they opened a storefront space in Northeast Minneapolis this past spring. By offering participants unmarked "flights" of water from different sources, such as various municipal water supplies, and asking people to pay attention to differences in taste and smell, the artist team generates greater discussion and awareness about water quality, water processing, and care of our precious resources. The project builds on ideas and experiences from Collaboratory field trips and discussions.

The variety and inventiveness of these projects is inspiring; they are a testament to what can happen when creative professionals have time to talk and dream together. As we embark on the next phase of the City Art Collaboratory, I am filled with anticipation for the new ideas, relationships, directions, projects, and experiments that will emerge in the coming years. This volume takes a snapshot of our first four years, and we hope it inspires others to consider this model of inventive incubation and hybridization.

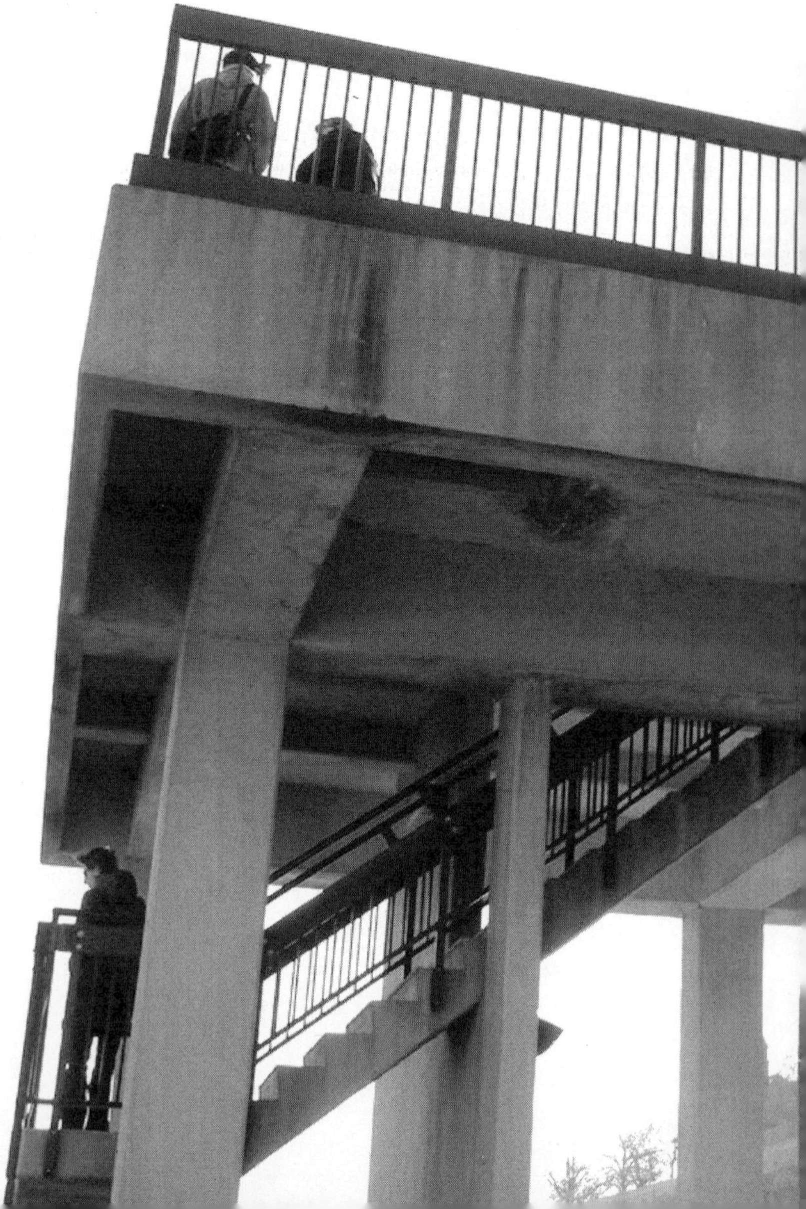

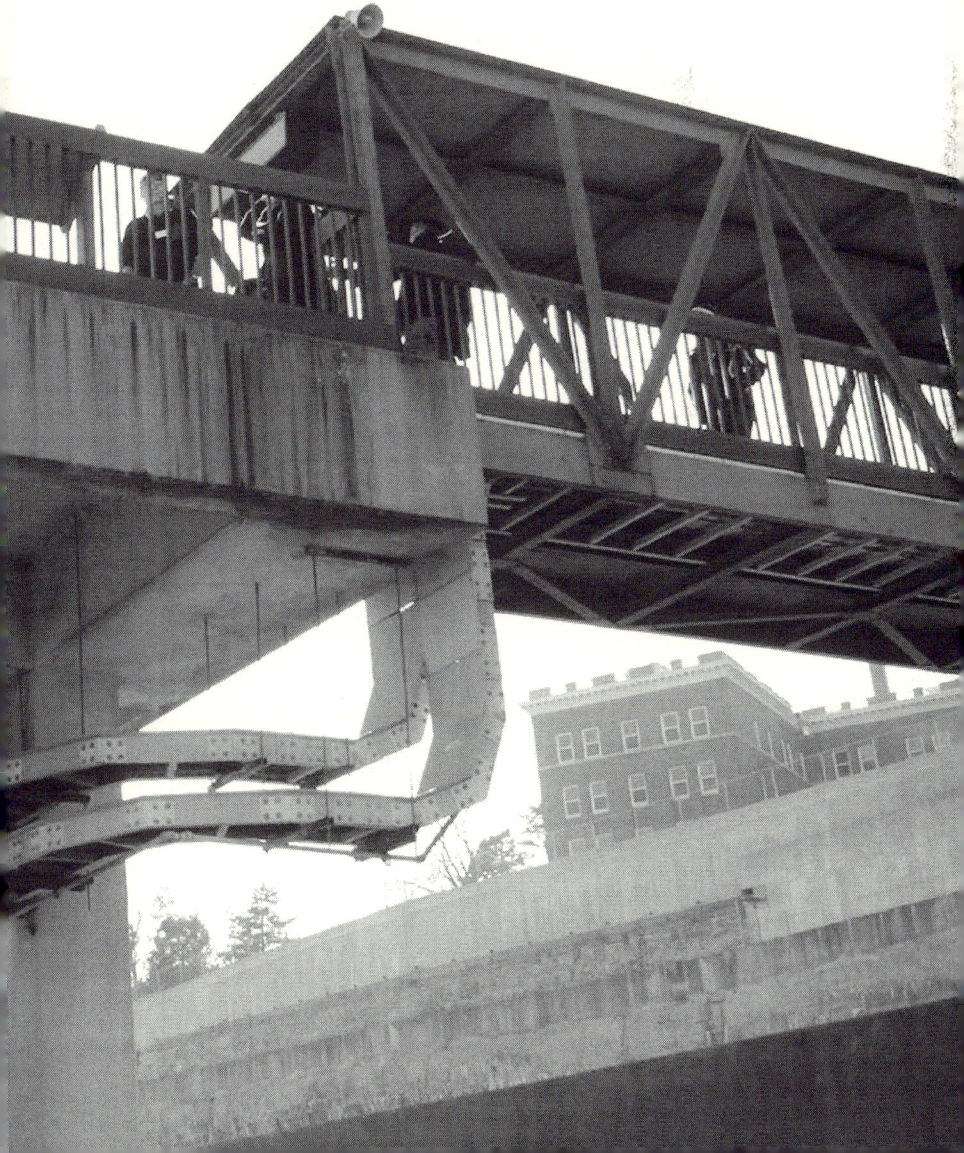

Preceding pages: The first Collaboratory field trip, a tour of Lock and Dam No. 1 (Ford Dam) in 2014. Photo: Andrea Steudel.

Meandering Methodologies

Shanai Matteson

Shanai Matteson became Director of the City Art Collaboratory in 2012 after developing the program with artists Olive Bieringa and Marcus Young. In addition to her role leading the program, Matteson is also an artist, and is now a frequent collaborator of many artists and scientists featured in this book.

Long before I became involved with Public Art Saint Paul through the City Art Collaboratory, the organization had begun experimenting at the intersection of art, science, and environment. Founder and then-President of Public Art Saint Paul, Christine Podas-Larson, reflects on the origins of these early efforts in her essay "Arising From the River" on page 245 in this volume.

> "Artists in the public sphere flourish when they collaborate with professionals across disciplines...In their own ways, [they] go forth in creative scientific and artistic action to have rippling impacts throughout their networks and communities."

In 2008, this spirit led Public Art Saint Paul to initiate the Sustainable Artmaking Fellowship, a program that offered $5,000 grants to artists for the purpose of researching the environmental impact of their artmaking, as well as setting goals to become more environmentally aware and sustainable.

For some artists in this initial fellowship group (which included Collaboratory Curator Olive Bieringa and current Public Art Saint Paul City Artist Aaron Dysart), this meant looking at the sourcing and disposal of art materials, such as ceramic glaze, or the woods and metals used in sculpture. For others, it meant calculating the carbon footprint of air travel around the globe for performances, or the hyper-local environmental impacts of place-based works. Each of the artists who took part in the Sustainable Artmaking Fellowship came away with a greater understanding of their impact on environmental systems, not only as artists, but as people living within ecological limits. They also came away with emerging theories about their role as environmental artists. Naturally, this led to further questions.

Through this first environmental-art fellowship, Marcus Young (then City Artist with Public Art Saint Paul) formed a collaborative relationship with dance and media artist Olive Bieringa of the BodyCartography Project. The two began talking about other scientific and environmental themes they hoped to incorporate into their lives and work, including things like behavioral science, climate change, and environmental activism. They concluded that one of the barriers they faced in their artistic development was making meaningful cross-disciplinary connections. They just didn't know very many scientists, and found it difficult to form collaborations with people who spoke such different languages and knew so little about one another's methods. After discussing this with Christine Podas-Larson, Hunting and Gathering Walks was launched.

The premise of Hunting and Gathering Walks was simple: it was a chance for artists and scientists to walk and talk with one another, to meander through place, and to become more familiar with one another's methodologies. By sharing experiences with place and comparing disciplinary perceptions and approaches, these artists and scientists were able to begin an exchange about the varied ways that questions, research, and creativity are fundamental to the practice of both art and science.

Though the framework for Hunting and Gathering Walks was straightforward, it elaborated a complex theory about how transformation in environmental thinking and change will happen: slowly, through changes in perception and practice. Artists and scientists applied to take part in this program, and were matched with others from different disciplines for walks along the Mississippi River. Participating artists and scientists also took canoe trips, explored watersheds, and spoke in public about their experiences. In the process of walking and talking, without map or agenda, questions emerged and were exchanged, and reciprocal relationships began to take root.

Today, numerous public programs, conferences, and events are planned with the explicit goal of connecting professionals across disciplines. The logic underscoring many of these efforts is one of increased productivity and the acceleration of innovation: the more diverse the people are within our network, the more information and perspective we have access to when we want it. While this may be true and important for any attempt to solve problems we've already identified, Hunting and Gathering Walks approached the value of relationships across disciplinary boundaries from a different vantage point. Rather than stressing the efficacy of "connecting" for the value of production or immediate problem-solving, this program was about slowing down, inviting participants to be themselves in the company of others, and fomenting transformative relationships.

Artist-led programs like Hunting and Gathering Walks are about affect rather than effect. The goal is not just to solve a problem or even answer a question, but to ask better questions. To arrive at better questions, we first have to embrace the curious, complex, and creative impulses of each individual, and allow for new understandings and insights to emerge from friction. Artists and scientists both work with ideas as well as tangible objects and experiences; in their own ways, they both produce meaning from and about the world around us. Hunting and Gathering Walks sought to establish an in-between space—albeit a temporary one—for artists and scientists to meet and develop their work and relationships. Over time, this has had the unintended impact of building a "community of practice," one that continues today through collaborations and friendships that began years ago.

At the close of Hunting and Gathering Walks, Public Art Saint Paul held a symposium with the curious title, "Artists, Scientists, and the Environment: What is possible together?" At that time, I was working through similar questions as a Public Program Curator at the University of Minnesota's Bell Museum of Natural History, where I had established another art-science project, the Resident Artist Research Program. I was invited to attend the Public Art Saint Paul symposium, a much-needed opportunity to further connect with a group of people that included artists, scientists, engineers, and environmental advocates.

The symposium began with remarks from artists Olive Bieringa of the BodyCartography Project and Marcus Young from Public Art Saint Paul. Invoking a history of artists and scientists working together on experimental projects, Bieringa and Young showed the group excerpts from the film *9 Evenings: Theatre and Engineering*, which documents performances curated in 1966 by Experiments in Art and Technology

(E.A.T.)—a project founded by artists Robert Rauschenberg and Robert Whitman, along with engineers Billy Kluver and Fred Waldhauer. Rather than facilitating art-engineering collaborations, E.A.T. instigated situations for open-ended contact. Artists were invited to connect with engineers from Bell Laboratories, developing collaborations as the creative process moved them, not according to the goals of Bell Labs. After an introduction to E.A.T. and other like-spirited examples, Young and Bieringa invited the 50 attendees of Public Art Saint Paul's symposium to break into smaller groups according to the conversations they felt needed to be explored.

One of those conversations stands out in my memory. It was framed by a question that led, in a meandering way, to the formation of the City Art Collaboratory. Marcus Young asked the group to consider, "What if we planned cities in 1000 year increments?"

This is exactly the kind of question biologists, geologists, and other environmental scientists at the symposium were eager to answer. It deals in scales of time that are familiar to those who look closely at evolution or geomorphology, as well as those who are moved by the ways cultural and spiritual practices evolve over time. When we talk about the future of cities—of things like development along the Mississippi Riverfront, or our efforts to mitigate stormwater pollution—we rarely look very far into the past or future. Perhaps we should.

When we think and act in cycles bound by economic development or the life of buildings and other infrastructure, we can easily forget that the Mississippi River didn't settle into its present course in our lifetime, or even within the lifespan of cities like Saint Paul. The river has changed dramatically over thousands of years, and it is still meandering, albeit slowly, in ways we may not perceive or easily comprehend.

Why does this kind of inquiry matter? Our small group discussed a "1000 Year Plan" with sincerity, and in the process, hit upon a truth that has had staggering consequences for our society and our environment: within capitalism, the prevailing paradigm that regulates the development of cities and systems, our memory is incredibly short. And because our memory is so short, our imagination about the future is also limited to what we believe we have the power to change in our own lifespan.

The conclusion of our discussion that day was an empowering one: we can decide, as artists and inhabitants of this place and time, to expand the horizon of our imaginations, and our imagined influence. We can choose to set our goals beyond a single project and it's measurable outcomes, even beyond our own careers or the lifespans of our organizations. We can think of ourselves as an evolving community of practice, one with generations and elders, attuned to the cycles of change that are all around us in the living and non-living elements of the places we inhabit. We can ask about the immediate tangible outcomes of our efforts, and look at the ways creative or collaborative acts ripple outward over time.

I left the Public Art Saint Paul symposium inspired, wondering what this shift in thinking might mean for those of us who create art, support artists, and care about the future of our places and ecosystems. Could we combine an understanding of the rippling impacts of collaborative works of art or science, with a deeper inquiry into the ways cities and cultures evolve over time? We don't always know the long-term effects of the actions we take today, but might proceeding slowly and with intention be a means of resistance? Can meandering be a slow way of changing the course of systems that are hurtling us toward an uninhabitable planet? And finally, how might this inform our ways of making art and supporting artists?

Following the symposium, Public Art Saint Paul made a decision to launch a new program that would build on the conversations and lessons learned from both the Sustainable Art Making Fellowships and Hunting and Gathering Walks. This program would have as its primary goal the development of relationships between and among a cohort of artists, science professionals, environmental advocates, and the City of Saint Paul, with the Mississippi River as a focal point. Marcus Young, Olive Bieringa, Christine Podas-Larson and myself, in conversation with other art and science professionals, created the framework for this new program. This is what we came to call the City Art Collaboratory: a process of inquiry and reflection, research and development, and the formation of new and deeper relationships.

We formed the Collaboratory by issuing an open call for participants. In their role as public art curators, Young and Bieringa selected a cohort of seven artists with very different professional backgrounds and artistic approaches. That group expanded in year two of the program. Among the Collaboratory Art Fellows were sculptors Aaron Dysart, Peter Morales, and Asia Ward (who also works as a science educator); puppeteer and playwright Janaki Ranpura; social practice artists Katie Hargrave and Molly Balcom Raleigh; digital projection artist Andrea Steudel; photographer and public artist Amanda Lovelee; and artist-designers Emily Stover and Aaron Marx. We intended the City Art Collaboratory to be a research and development space for these artists, who all shared an interest in working in the realms of nature, urban design-development, and environmental sustainability.

Science and engineering professionals from local universities and colleges, public agencies, nonprofits, and private companies were invited to participate through a separate but similar open call. After selecting artists, we asked those artists to review the applications of science professionals, and to tell us which among the applicants they were most interested in connecting with, based on their own artistic development goals.

Our final cohort of Collaboratory Science Fellows included graduate-level academic researchers Ilana Percher (Physics, University of Minnesota), Daniel Nidzgorski (Ecology, University of Minnesota) and Adam Kokotovich (Natural Resource Policy, University of Minnesota); researcher Kate Brauman (Institute on the Environment, University of Minnesota); academic researchers and professors Jerald Dosch (Ecology, Macalester College), Roopali Phadke (Environmental Studies, Macalester College), John Schade (Ecology, Saint Olaf College) and Kirsten Valentine Cadieux (Environmental Studies, University of Minnesota and now Hamline University); science policy and communication professional Leili Fatehi (Humphrey Institute and now Apparatus Public Affairs); engineers Bunmi Odumuye (Renewable Research and Policy Institute) and Matt Kumka (Barr Engineering); and environmental educator Leslie Yetka (Minnehaha Creek Watershed District, and now Freshwater Society).

Before we began, we asked everyone involved in the Collaboratory to tell us about their goals for learning and for connecting across disciplines. What did they already know, and what did they want to know? What was their commitment to and interest in interdisciplinary collaboration?

That was in 2012. Now, four years later, this book is an opportunity for a deeper reflection, both by City Art Collaboratory organizers and participants, and by critics who spent time with our group. It is, like any publication anchored to a specific moment in time, incomplete. But our hope is that it might offer insight to those launching similar interdisciplinary projects and organizations, specifically those involving the critical contributions of both artists and scientists. In an era marked by the intertwining ecological and social impacts of climate change, to name just one urgent example, relationship-building across disciplines is not only a matter of vision or integrity, it may be our best hope of evolving our society in the direction of a habitable future.

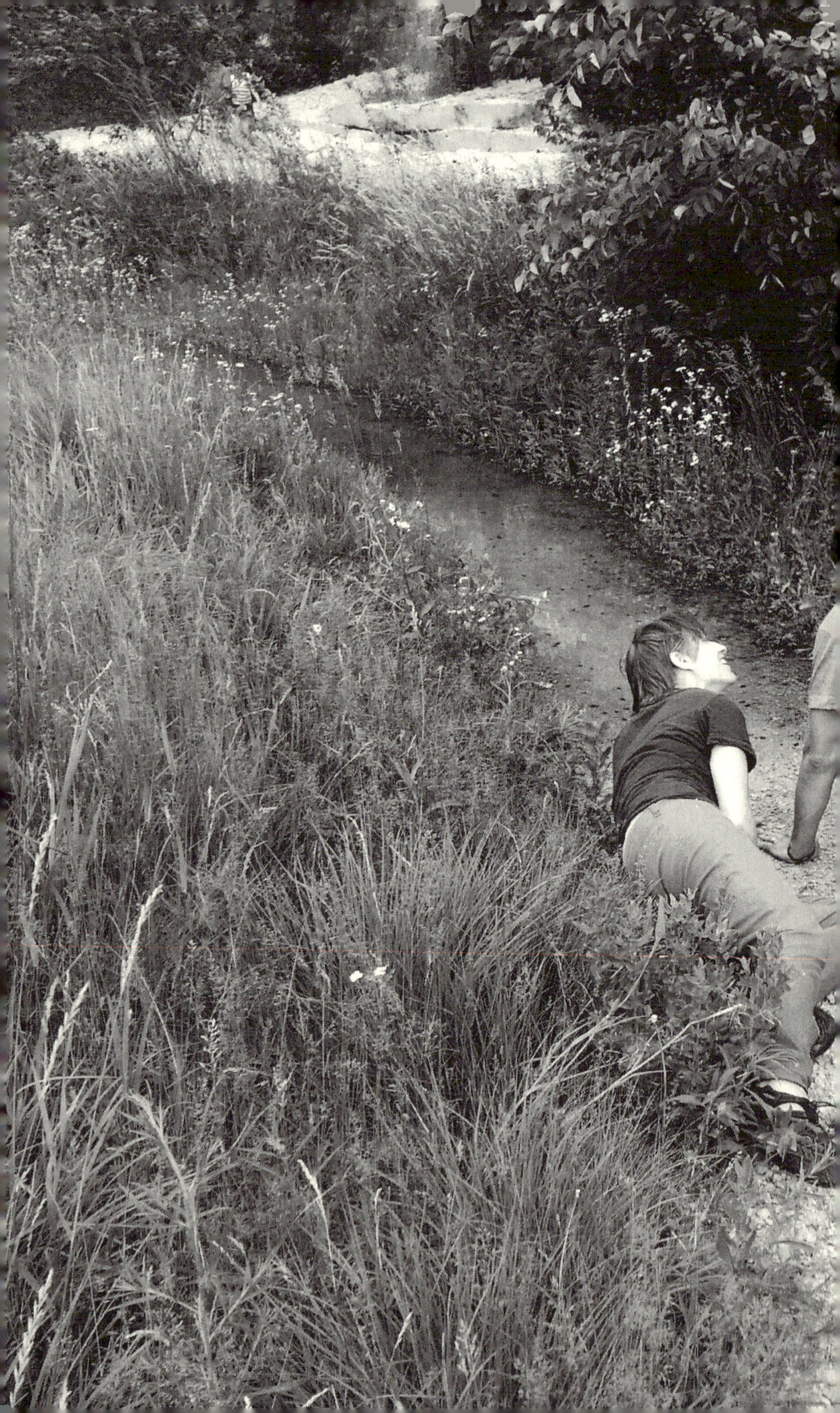

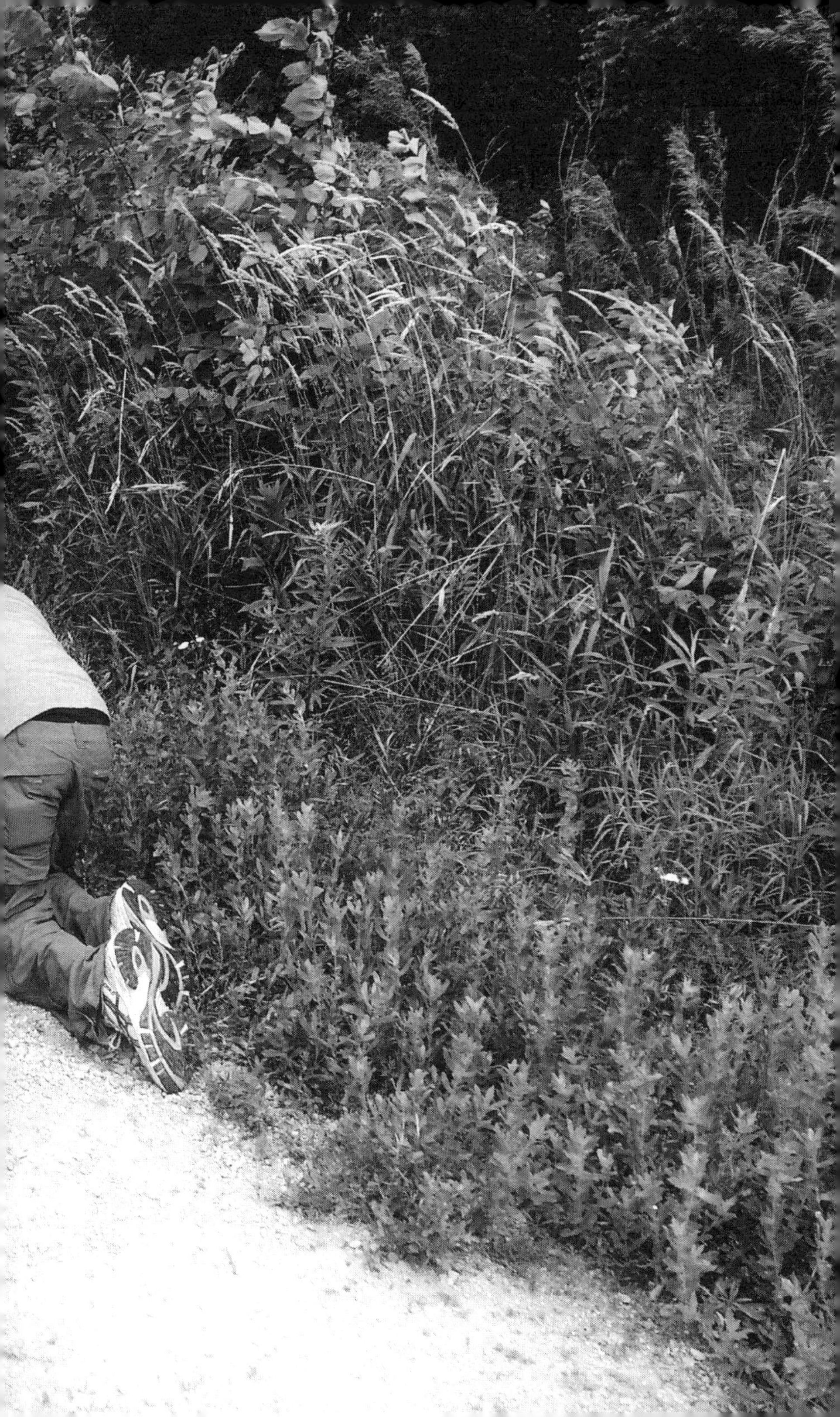

Preceding pages: Collaboratory Art Fellow Aaron Marx and Collaboratory Curator Olive Bieringa (the BodyCartography Project) on a field trip designed and led by Olive Bieringa at Bruce Vento Nature Sanctuary in 2014. Photo: Andrea Steudel.

Interview: Olive Bieringa of the BodyCartography Project

Olive Bieringa with Shanai Matteson

Olive Bieringa creates work with Otto Ramstad as the BodyCartography Project. The mission of BodyCartography Project is to engage with the vital materiality of our bodies, creating dance in urban, domestic, wild, and social landscapes.

As one of two artists to lead Public Art Saint Paul's Hunting and Gathering Walks, Olive Bieringa brought her own experiences with art and collaboration to the development of Public Art Saint Paul's Art and Environment Programs. Along with artist Marcus Young, Bieringa developed the initial concept for an art and science fellowship. She also helped to select Collaboratory participants from an open application process as one of two City Art Collaboratory curators. In this short interview, Bieringa reflects on the origins of her interest in art-science collaborations, as well as some of the ways her participation in the Collaboratory shaped subsequent projects she's developed through the BodyCartography Project.

Shanai Matteson: *Can you talk about the evolution of your thinking with regard to art and science collaborations, especially those with environmental implications?*

Olive Bieringa: My interest, as a dancemaker and somatic movement therapist, was at first in trying to understand the relevance of my own work, how it could intersect or have an engagement with other disciplines. How could work of a physical nature have an impact on the way we think and feel about climate change, for example. That was very much my early history. I made a solo piece that investigated this theme with an enormous block of ice in 2006, a piece called *Condition.*

SM: *What was the reaction of the community of artists you were involved with at that time?*

OB: At the time I was having conversations about interdisciplinary practices with a friend of mine, Margit Galanter, who was working at Earthdance, which is a dance retreat center in Massachusetts. We were interested in how we could create a space for that conversation in the dance community, which resulted in a festival called SEEDS—Somatic Experiments in Earth, Dance, and Science—which happened from 2008 - 2010 and is running again in 2016.

SEEDS brought together artists working in many different disciplines with environmental educators, scientists, and permaculture practitioners. There were people making installation art as well as dance. Over the three summers we tried many different kinds of set-ups: workshops, walks in the woods with scientists, and having artists and scientists hold a public conversation. Everyone else was there to try out ideas, or to listen, or to ask questions. We tried a residency model, commissioned structures, created permaculture ponds, and a twig tea house. We built structures for mushroom spores, dug up a spring, and facilitated a lot of performance work. There was a lot happening between somatic practices, like Body-Mind Centering®, coming out the materiality of the body. We were interested in systems thinking, and moving between the systems of the body and earth.

It's funny, because right now we're in such a different moment—you know what I mean?

SM: *In terms of languages around these practices?*

OB: Yes. It seems like design thinking has evolved to bring art and science together in a very particular way, and now you have this whole community engagement piece, as well as conversations about leadership in cross-disciplinary work. I feel like it's what's coming up for me right now, in terms of the visibility of those two things. It's almost become an agenda: leadership, engagement, and design thinking, as a way to develop that intersection between art and science. And then social practice is sitting over here to the side, also part of the conversation.

SM: *How did you start working with Public Art Saint Paul?*

OB: I got into the conversation with Public Art Saint Paul as a Sustainable Artmaking Fellow in 2008; the BodyCartography Project got that fellowship. We were there to explore ways we could be less consumptive within our own work, but this also had an effect on the content of our work. Since 2005 we'd already been developing a large piece called *1/2 Life*, which was focused on nuclear issues around the Pacific from the perspective of Japan, New Zealand, and the U.S., which premiered in 2010.

Out of the Sustainable Artmaking Fellowship, Hunting and Gathering Walks was developed, which was similar to a model that had been proposed at SEEDS by Melinda Buckwalter. Artists and scientists submitted proposals and we then selected a number of them. Selected artists or scientists could invite each other to go on an a walk and talk along the Mississippi River. I took a canoe trip down the river with one scientist, and I walked along the river with someone else. I went to the Saint Anthony Falls Research Lab and I walked through a watershed with ecologist John Schade, just to observe and experience what a watershed meant.

And as part of that project, we held three public talks, Q&As with artists and scientists, as well as a day-long open space event , which is where you came in, Shanai.

SM: *You mentioned that when you began thinking about these intersections, it was a different time. Do you feel that the idea of interdisciplinary collaboration is more accepted today, or do you feel that the emphasis is now on something else?*

OB: Interdisciplinary practice now seems much more common. Look at the d.school at Stanford for example. Places have emerged in which ideas are being seeded, and these ideas are coming out of exchanges that are explicitly cross-disciplinary. I think that's something we were hungering for with the artist and scientist walks. It wasn't even like we needed to find a project, it was just that being in exchange was very stimulating for the brain, trying to understand someone else's discipline or field, and what is it they engage with day-to-day. In turn, how do I understand that through my own practice, and what can my practice bring to the conversation? I think that is happening in a much more widespread way than it was eight years ago.

SM: *Do you feel like there is more emphasis on a project now? Do people have an agenda, and what are they trying to get out of it?*

OB: I don't think it's project driven, I think that now creativity is well-recognized. There is value in creative approaches. I think it's really shifted. This interdisciplinary work is what we're doing, and we're doing it because it's useful, not just because of a single project, but the myriad outcomes that the process could generate—even just what it does to the mind. Change only comes about by changing one's practices, not by simply changing the way you talk about what you do.

SM: *Yes, I would say that in the past few years I've heard a change in the way people talk about interdisciplinary work, from "It's interesting," to "It's useful," and now to "It's necessary." Problems suddenly seem so big that you can't even attempt to address them without interdisciplinary thinking and collaboration. You need all of those ways of thinking, seeing, and doing in the mix.*

OB: Yeah, so it seems that in a way, you need the thinking of social scientists, hard science research, artists, indigenous practitioners, designers, and community engagement practitioners, where everyone is playing a role in the conversation.

SM: *How have the experiences that you've had through projects like SEEDS, Hunting and Gathering, and the Collaboratory changed your artistic practice and work?*

OB: I definitely feel that there's something about the activity of trying to understand someone else, their language and practice, that makes your brain engage in active problem solving. If you stay in your own field, and you're working with the jargon of your own field, it becomes very self-referential. You can get into amazing detail and complexity, but somehow I really like that feeling of trying to understand a different way of working from outside of my field completely.

I feel like I'm being stretched, even if I can't solve a problem, or can't fit the puzzle pieces together. Somehow the very act of being in this place of challenge makes me think differently about my own work, and puts my own problems in a different perspective. It also makes me feel more playful with what could be possible.

SM: *What does that look like for you, as an artist?*

OB: I think because I work in the practice of dance, I'm working in a relational field, not in isolation. I make work with many collaborators; dancers, musicians, designers, audiences. I have been working in public spaces for eighteen years bringing dynamic physical practices, alternative movement behaviors and engaged embodiment. I work with the materials of the body. I constantly practice coming into relationship with the world we are working in. I practice being present. Presence is about being in relationship, with oneself, one's cells, one's environment, and other beings. This capacity to fluidly be in relationship amidst many shifting conditions, requires a quality of porousness and deep self-knowing.

The felt experiences of our bodies are key to understanding our interconnectedness with others bodies and other life forms. We've really started making work that functions, in many ways, within the realm of social science. I'm no longer interested in something that is site-specific, I'm interested in something that is context-specific, and has a social function. The work we make brings its own set of problems and challenges, but it also needs people in order to work. It needs a public, it is not just an object being created for itself. Even in the process of rehearsing these new projects we need to do prototypes all the time, and we have test audiences coming in all the time in order to understand what it is we're making.

Our work is located in the relationship between the performer and the audience, in the bodies of the performer and the audience, in the memory of the performer and the audience, and our personal interpretations of those experiences.

SM: *So it's like a design thinking approach to dance?*

OB: Right, and I'm working with a graphic design team, a social worker, and someone who is developing an online container that will be the interstitial fluid between all other aspects of the project. We're also working differently with our dancers. They each have a production role as well, such as music composer, costume designer, or other aspect of responsibility.

The other change is that these new works have room for each person to develop and evolve. We have a social responsibility to the people within the work, as well as the public who is coming out to engage. In a way, we're trying to use dance to reveal an awesome way of being together, and how to create a structure that encourages us to function the best way we can as humans, to bring out the best in people. So the whole thing is really built with a spirit of generosity that is sustainable. It's not like wasting tons of resources on something that is just a vanity project or a one-off. We're creating an experience that is focused on transformation and change.

SM: *What do you think about the future of these transformative processes—of making a dance as demonstration, or the Collaboratory—which it seems is also a process of individual and collective transformation, by changing the ways we inhabit place and form relationships.*

OB: What the Collaboratory offered that was so important for me was a way to understand different systems of the city, and some of the political structures that are functioning alongside these different systems that we know so little about. This conversation about resources, like water for example, many of us take for granted. All of those experiences gave me the opportunity to connect with people I could not have connected with on my own. And there is something about the possibility for those connections to be the great refresh we need to further our lives and work.

SM: *I certainly see things that way, but I know it can be hard to make a case for how important that is. We've talked about building relationships and changing systems being more of a matter of following an arc over a longer period of time—would you say that's the case for you?*

OB: Yes, and sometimes you don't see the arc until afterward. Sometimes these things happen and we talk about immediate context, and the choices you need to make, but then new opportunities arise, or challenges, and we forget what had set us on this course. Those earlier experiences really do shape our trajectory. I think our understanding deepens, and we go from metaphor to something we're trying to reach for. Eventually, it becomes embodied, and then it becomes systemic. That is what makes the arc.

SM: *So the practices that eventually change your life and those systems begin as singular experiences?*

OB: Exactly. One example of that is this motion of rocking from side to side and singing. This has become a part of our choreographic material, and that began to happen around the time our daughter was born. I mean, how many hours do you spend just swaying, holding your child? It's not like we sat down to plan out what kinds of movements we were going to make in order to become parents, to care for our children. But then, here we are.

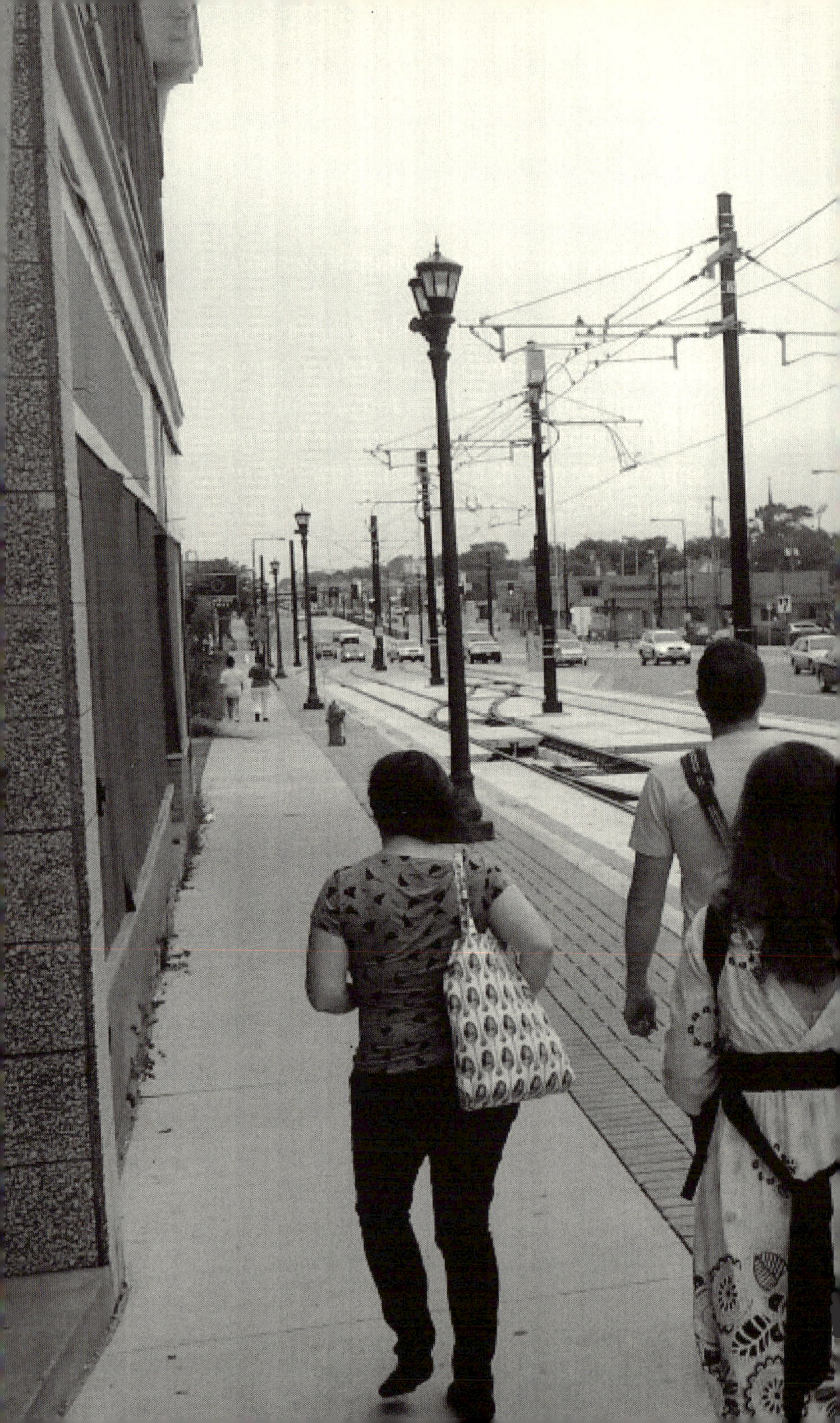

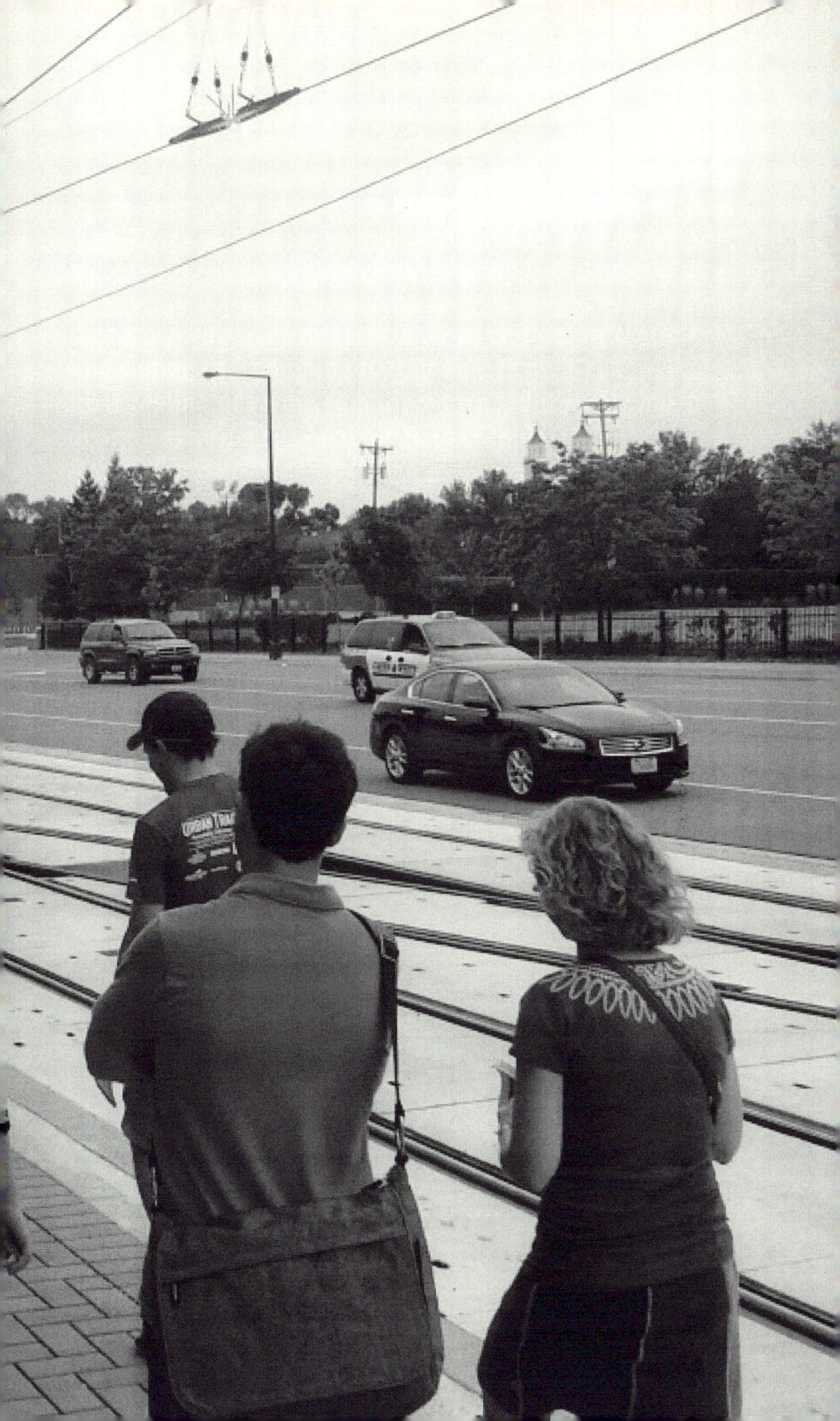

Preceding pages: A Collaboratory field trip along the Green Line, a light rail corridor connecting downtown Saint Paul with Minneapolis, led by Collaboratory Science Fellow Matt Kumka and his Barr Engineering colleagues in 2014. Photo: Andrea Steudel.

A Retroactive Manifesto

City Art Collaboratory

In the spring of 2015, with the help of writer and educator Christina Schmid, the City Art Collaboratory gathered to articulate a set of shared values emerging from Collaboratory activities and conversations.

What is City Art Collaboratory?

We are a cross-disciplinary cohort of public artists, architects, scientists, educators, engineers, and environmental advocates who share an insatiable curiosity and desire to collaborate on publicly-oriented, environmentally-attuned projects that change the way cities are built, experienced, and sustained.

Convened and supported by the non-profit organization Public Art Saint Paul since 2012, we continue to develop and evolve through a process of individual and collective inquiry, direct experience, reflection, and ongoing research-development.

What do we do?

Together, we experience, engage, and ultimately disrupt—with joy and urgency—the vital, complex, and often obscured systems that comprise life in our city. We explore ecological, physical, political, and social systems in conversation with the people who design and maintain them, through open-ended field trips, workshops, and our own public art, science, and engagement projects.

Why do we do what we do?

We seek to identify and illuminate persistent problems, opportunities for engagement, and leverage points for systemic change that transcend individual fields of concern.

By making this process visible and relevant to wider publics, we also seek to amplify curiosity, to surface hidden or overlooked systems and processes, and to expand knowledge of and access to opportunities that meaningfully shape the way cities are built, experienced, and sustained.

How do we do it?

We create and value the time and space necessary for direct experience with place and process; for sustained reflection; for creative experimentation, risk, and failure; for the slow and thoughtful development of individual practitioners, collective creative processes, and the ideas and projects they develop.

We believe that this approach will have a collective impact far greater than approaches that begin by framing problems and pursuing solutions from where we are. This approach does not begin or end with what we know or can do, but rather, with our most urgent questions and the unexpected possibilities that come from pursuing these questions together.

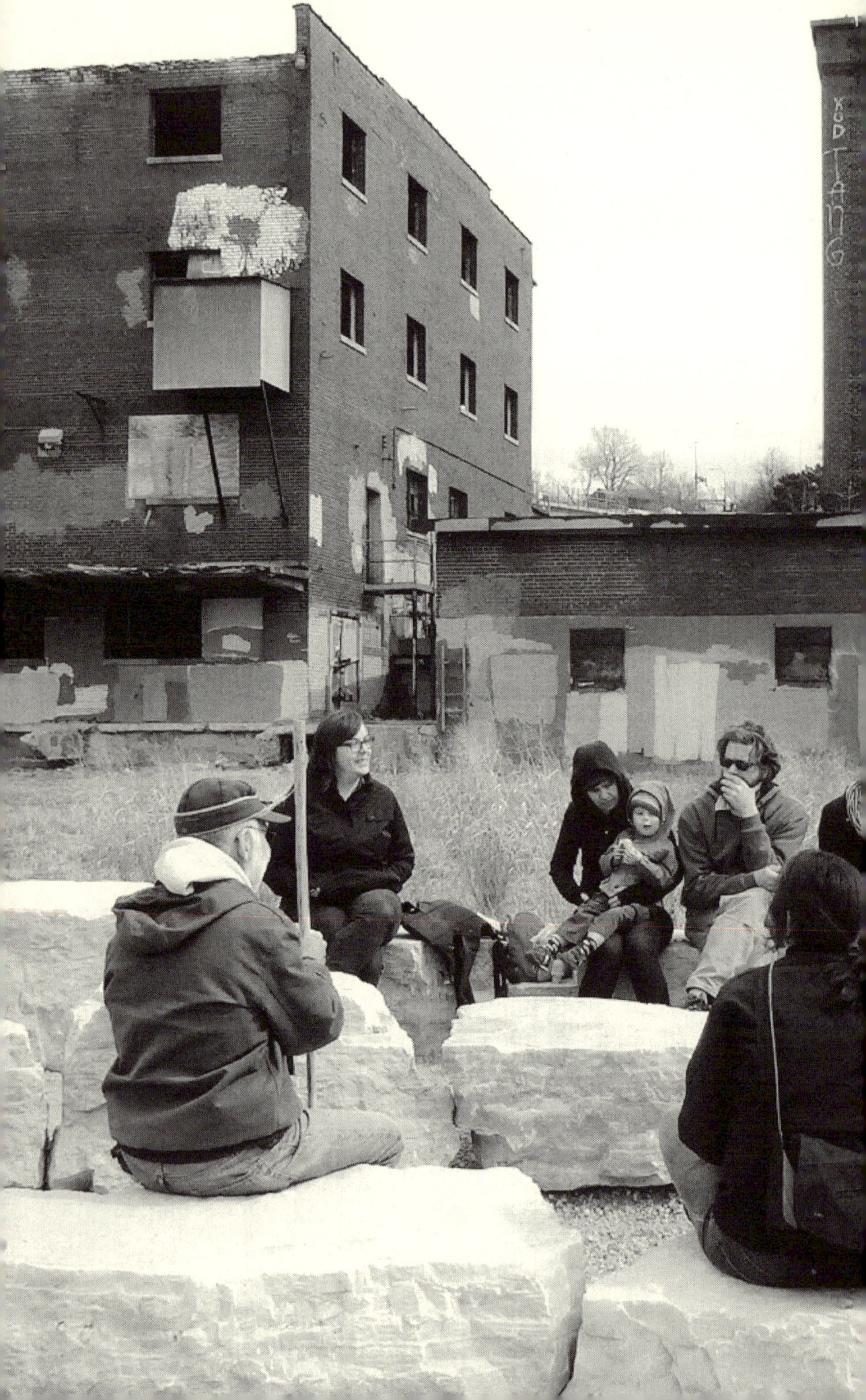

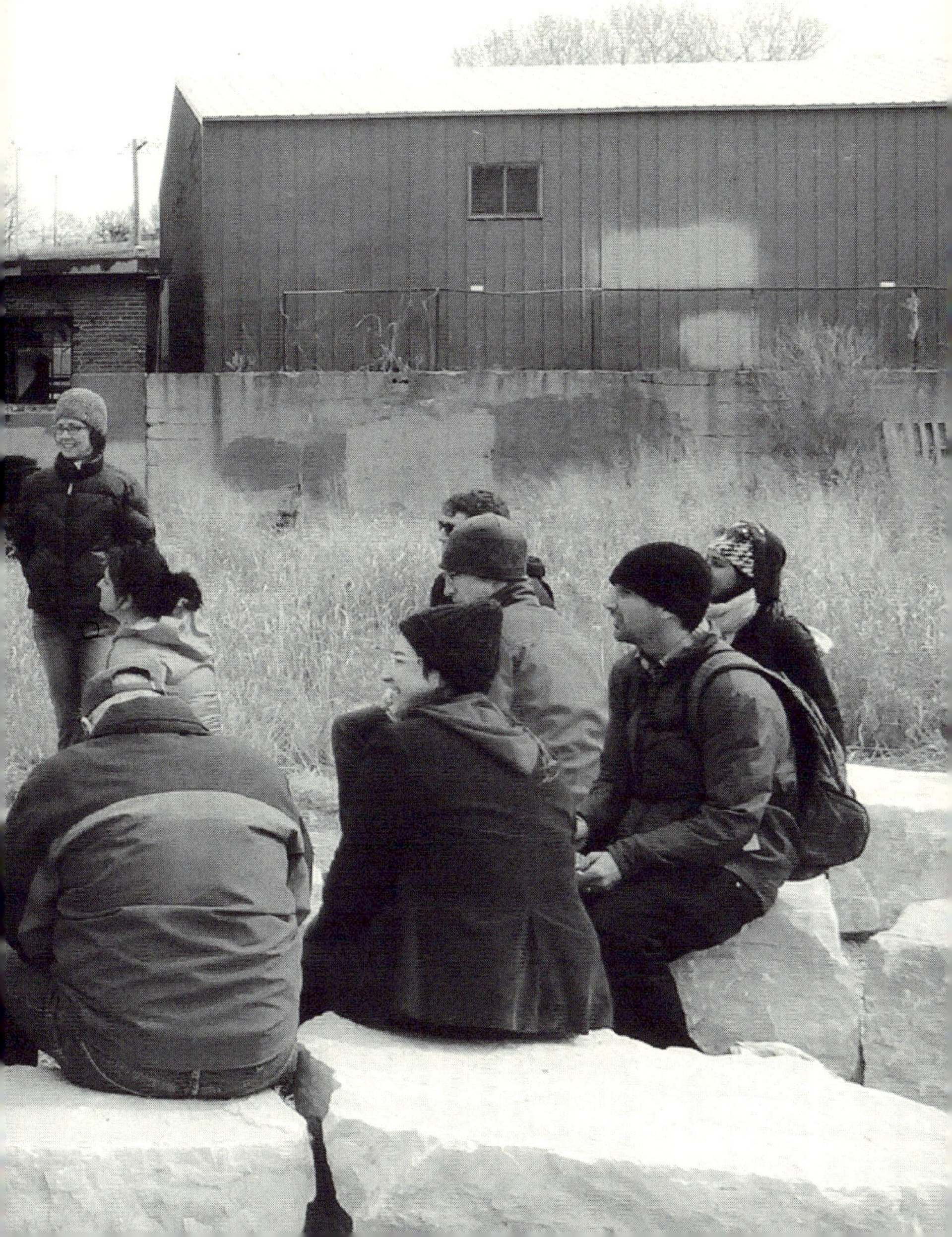

Preceding pages: A Collaboratory field trip to Bruce Vento Nature Sanctuary in 2014. Photo: Andrea Steudel.

Art in Place: The City Art Collaboratory

Christina Schmid

In 2015, writer and educator Christina Schmid was invited by the City Art Collaboratory to lead a workshop about artist statements and manifestos. After spending time with the Collaboratory cohort, she authored this essay.

In 2012, when a group of artists and scientists came together to form the first cohort of the City Art Collaboratory, little was set about the group's purpose or direction. Public Art Saint Paul initiated the Collaboratory program, intended to facilitate conversations across disciplines and to generate new perspectives at the intersection of creative problem solving and scientific methods. From the start, process mattered more than predetermined outcomes. Unfettered by disciplinarity, the City Art Collaboratory was conceived as an incubator for speculative, interrogative, and unorthodox inquiries. Despite its openness, some shared values guided the group from the start: a passionate interest in ecologies, creative thinking around issues ranging from water quality to civic service and community engagement, and a commitment to place. The City Art Collaboratory's emphasis on locating its cross/anti-disciplinary investigations—dubbed "deviant interdisciplinarity" by one of its members—in the city of Saint Paul is the focus of this essay.

In conversations with members of the City Art Collaboratory, place, understood as the multifaceted nexus of interlocking systems, quickly emerges as a crucial idea. While some systemic phenomena occur based on the geology, seasons, and climate of the place, others are part of an urban infrastructure, created and administered by human agency. Studying the former yields information to implement the latter, which in turn affects the material conditions of a given site. From this point of view, place becomes a multi-tiered, complex feedback loop. On field trips, members of the City Art Collaboratory studied how key points in such site-specific infrastructures support public health and allow urban life to function. In Saint Paul, the practice included visiting a wastewater treatment plant, the city's snowplowing and road salting operation, and a heating and cooling plant on the banks of the Mississippi River.

When they work well, these infrastructures often remain invisible to the public eye. But making the "plumbing of the city" visible, as one Collaboratory member put it, is part of a shared commitment to understanding the city of Saint Paul from a wide range of vantage points. Thus, the City Art Collaboratory's approach differs from other creative organizations such as the Center for Land Use Interpretation, based in Los Angeles, whose primary goals are documentary: making visible and bringing into awareness the ways we, as a culture, use land. Rather than serve documentary purposes, the City Art Collaboratory's field trips provide opportunities to see the city as both problem to be solved and, through its administrative structures, as problem solver; to ask questions, not as fellow experts and administrators, but as artists and scientists; and to practice an understanding of art and science as embedded in and responsive to the challenges a particular community faces. This explicit focus on place and locatedness runs counter to the trends and tendencies in much of contemporary art.

When did place become suspect? In most contemporary art circles, place has a bad name. Ever since "creative place-making" became code for gentrification and economic development, place has been considered passé as a subject. Place connotes roots, which suggest something sedentary, or simply unsavory in an art world where rootlessness rules. Mobility matters. Place is considered by many to be provincial at best, cliché and kitsch, shackling artists and their work to a localism at odds with the global aspirations of the art market. For many artists, itinerancy has emerged as the new paradigm. Simply put, who moves matters. Success means having the means to travel. Artists are cast as "semionauts," travelers who manipulate signs, cosmopolitan trendsetters, nomads of a new age, where being in transit offers a curious freedom from lasting commitments and responsibilities. This paradigm prizes the ability to plug into contexts, complete a project, and plug out. The transience of such peripatetic practices equals commitments for a limited time only. Anything else is suspected of regionalism, localism, provincialism—of being tied to a place, confined by its particularities.[1] Yet what may seem anachronistic and limiting at first glance ultimately reveals the risks harbored by today's globally networked art system.

In Minnesota, ambivalence concerning place runs deep. On the one hand, the allure of a more cosmopolitan elsewhere is palpable, and frequently accompanied by a kind of upper-Midwestern inferiority complex. In order to matter, art has to come from the coasts, this mindset says. On the other hand, localism as "pride in place" flourishes where annual per capita arts spending, the non-profit art culture of the state, and the 2008 Legacy Amendment are concerned—all clear signals that the voters of the state value art and their environment. But recent re-branding efforts of the region, proposing "the North" as a new signifier of Minnesota's unique contributions to

1 For discussions of this new paradigm of mobility and its implications, see for instance, Nicolas Bourriaud, *The Radicant* (New York: Lukas & Sternberg, 2009), Bartholomew Ryan, "If You Are Willing: The Army of Individuals" (New York: D.A.P./Distributed Art Publishers, *9 Artists*, 2013) and Lane Relyea, *Your Everyday Art World* (Cambridge, MA: The MIT Press, 2013).

U.S. culture, attest to the lingering discontent with being stereotyped as "fly-over" country.[2] The City Art Collaboratory counters these deep-seated reservations by its clear commitment to studying and engaging with what is right here, what may have been overlooked, what may indeed make this place unique. It examines the significance of place for creative practices that transcend current models of contemporary art. Participants cite service as a motivating factor for their practices and understand place as a platform for far-reaching exchanges, built on an intimate familiarity with place and people.

While the emphasis on studying one place may run counter to more transposable art practices, the City Art Collaboratory shares its open-ended and process-based approach with a host of other kinds of public and participatory art and socially engaged practices. A slew of recent publications and exhibitions demonstrates this international phenomenon: Suzanne Lacy, whose work has inhabited the space between activism and art for decades, published *Leaving Art*, an anthology of her writing, in 2010; Pablo Helguera's *Education for Socially Engaged Art*

2 See for instance Kevin Obsatz, "Cinema of the North," Minnesota Playlist, January 2015 (http://minnesotaplaylist.com/magazine/article/2015/cinema-of-the-north) or the Walker Art Center's panel discussion in November 2014 "Midwest? The Past, Present, and Future of Minnesota's Identity." (http://www.walkerart.org/calendar/2014/past-present-and-future-minnesotas-identity)

came out in 2011, Grant Kester's *The One and The Many* in 2011, followed by *Conversation Pieces* in 2013. Claire Bishop's rigorous study of participatory art, *Artificial Hells*, entered public discourse in 2012; Lane Relyea's *Your Everyday Art World* offered incisive analysis of today's art world in 2014. National conferences such as Creative Time, Open Engagement, and Hand-in-Glove 2015 (hosted at the Soap Factory in Minneapolis) attest to the widespread interest in socially engaged practices. MFA programs at a number of international art schools and universities prepare artists for work in "post-studio practices," engaged art, and relational aesthetics. Exhibitions, from Mary Jane Jacob's *A Proximity of Consciousness: Art and Social Action* along with the symposium, "A Lived Practice," in 2014 at the School of the Art Institute of Chicago to Creative Time's 2011 *Living as Form* (which also traveled to Plains Art Museum in Fargo, led at the time by Colleen Sheehy, now Executive Director of Public Art Saint Paul), from Carolyn Christov-Bakargiev's show at *documenta 13* in Kassel, Germany to the Walker Art Center's three-year summer program, Open Field in Minneapolis, to the University of Minnesota's *Thinking Making Living* and Public Art Saint Paul's renowned City Artist program, art that involves both social engagement and lived practice often appears at the cutting edge of contemporary art. But while the rhetoric surrounding socially engaged practices is on the rise, practitioners are often the first to point out the way of making work—in dialogue with others, invested in process rather than artist-determined outcomes—is anything but new.[3]

For Lane Relyea, the recent rise to international prominence of socially engaged, mobile practices marks a paradigm shift in the world of art. While Relyea maintains a profound ambivalence about the nature of this shift—an unease that hinges on the appropriation of creative DIY attitudes by neoliberal market forces—he nonetheless understands the changes that have already occurred in artistic practices as non-reversible. Since art never exists in a vacuum, 21st-century network culture, complete with platforms, databases, and information flows, has left its mark on the art world. Yet Relyea's measured reservations about the commodification of socially engaged practices are overshadowed by much stronger and widely received arguments against "critical art" by French philosopher Jacques Rancière.

In Rancière's substantial oeuvre, displacement, distance, and detachment play a key role in establishing the very efficacy of art. Here, distance changes from lack of attachment to any one place to a philosophical premise of art. Originally a scholar of the 18th century, Rancière traces what he considers art's aesthetic efficacy to the social upheavals in European societies. Art,

3 J. Morgan Puett, at her lecture in the Art Department at the University of Minnesota on November 19, 2014, emphasized that the practice precedes the current rhetoric by decades.

he argues, suffered a profound "loss of destination" when artistic sensibilities were no longer perceived as limited to a select, privileged few. Art thus extended its reach in the wake of the revolutionary fervor that swept the continent. No longer limited to any one audience, art could now reach anyone. Museums embodied this unprecedented placelessness: in the remove offered by such institutions, anyone could partake in the aesthetic experiences afforded by art. No longer limited by matters of class and taste, art could putatively speak to anyone. New patrons gained access to unfamiliar passions that redrew the very topography of the thinkable, feasible, and doable. According to Rancière, engineering such shifts in perception is the domain of art, not to be confused with the disruptions to common sense that occur in the realm of politics. In the collapse of the two spheres, the aesthetic and the political, Rancière locates critical art's profound misunderstanding of its own role in the world. Art's quasi-democratic openness came with a price: art had to remain separate, distant, detached, its politics only possible under the mantle of aesthetic distance.

Much of Rancière's disdain for critical art hinges on what he considers its flawed assumption that aesthetic effects can be predicted: if I show you an image of human suffering, you will have an emotional reaction that will lead you to want to end said suffering. If I show you an image of consumerism run amok, you will understand my critique and oppose the excesses of consumerism in your personal choices. Art culpable of such simplistic cause-and-effect calculations profoundly misunderstands what art actually can do in the world, in Rancière's mind. Socially engaged practices, rooted in place and committed to addressing concerns shared by a community, are also suspect in this philosophical framework: eager to assert its capacity to do more in the world than hang on a wall in a place-less place and speak to who-ever, to embrace interdisciplinary methods and means, socially engaged and civic-minded art confidently abandons aesthetic distance.

Abandoning such distance, though, does not equate to insisting on a predictable outcome. Indeed, as the City Art Collaboratory demonstrates, process may matter more than product. In this respect, the Collaboratory differs from other artists and collectives working at the intersection of art and science. Natalie Jeremijenko's lab at New York University, for instance—often known to adopt quirky and unorthodox approaches to interface with non-human species and address ecological challenges in the city—aims to produce tangible interventions, prototypes, and protocols of engagement that remain rooted in a purposeful, outcome-oriented framework. Process is ultimately valued for the results it may produce. Humor and metaphor aside, many of Jeremijenko's projects stay true to a scientific telos. In contrast, the City Art Collaboratory prizes the open-endedness and unpredictability of the shared conversations and field trips of artists and scientists. The process hopes to cultivate exchanges and learning beyond and across disciplinary boundaries, reshuffling the topography of what art and science can be and do in the world.

The benefits of art's loss of destination, applauded by Rancière as a foundational premise of aesthetic efficacy, are equally questionable. Distance and detachment ring hollow in 2016 when the streets of U.S. cities are burning, again, when protesters need to remind the powers that be that Black Lives Matter. Artists have long taken a stance against the destination-free, place-less sort of art Rancière (and Kant before him) advocates: notably, Adrian Piper and David Hammons, both artists long familiar with institutional racism, have taken issue with the presumed placelessness of art. Piper, in her still eye-opening treatise "Power Relations in the Artworld" (1983) dismantles the fake neutrality of formalism and demonstrates the values and ideology at the core of protestations in favor of "art for art's sake." There simply is no neutral ground to stand on. For Piper, knowing her audience and having an audience in mind is part of her artistic integrity, part of the perlocutionary force of her work. Hammons, whose departure from the gallery system coincided with his desire to make work for the community he lived in and with the place he knew intimately in mind, took creative action against the perceived arrogance of artists disregarding the human beings asked to interact with their artwork.[4] Both Piper and Hammons serve

4 In 1981, Hammons performed *Pissed Off* at the Federal Plaza

as examples here of artists whose lived experience led them to disavow pretensions of aesthetic distance.

Experience, then, affords us access to a kind of place; a cultural, social, and political position. Feminist theory addresses this sense of locatedness as positionality: we cannot help but be positioned in the intersections of race, class, gender, age, and ability. (And of course the list could go on). Not all such positions carry equal weight. Indeed, the farther removed one is from the paradigmatic center of social life, the more clearly one can see the mechanics of privilege in action. Once again, there is no neutral ground, no place-less outside in this web of socially relevant meanings.[5] Who can afford to lose sight of destination and cultural position of the work? When we lose a sense of place, we risk losing awareness of social location and position.

Remove and detachment sound similarly problematic in an age when drought lays waste to arable land and invasive species turn river ecologies upside down. If we want art to matter, to weigh in on such pressing concerns, formalism's distance seems profoundly outdated. Artists can no longer afford to adopt what Timothy Morton calls "beautiful soul syndrome": "the

---

in lower Manhattan, where Richard Serra's controversial sculpture *Tilted Arc* was then located.

5 Naomi Scheman, "Queering the Center by Centering the Queer" in Feminists Rethink the Self, ed. Diana Tietjens Meyers (Boulder: Westview Press, 1997), 124-165.

beautiful soul washes his or her hands of the corrupt world, refusing to admit how in this very abstemiousness and distaste he or she participates in the creation of that world. The world-weary soul holds all beliefs and ideas at a distance."[6] But for Morton, the time for artists and thinkers "to sit smugly outside the area" they are examining is over. Any genuine theoretical approach must mix thoroughly with the subject of inquiry: "The only ethical option is to muck in." There is no "away;" there never has been.[7] Without the illusion of a "safety net of distance," we no longer can pretend to "extricate ourselves."[8] The City Art Collaboratory, with its twin focus on service and ecology, never set out to operate from a distance. The foundational idea behind bridging the divide between art and science is to learn from such proximity of disciplines and histories; to understand communities, ecologies, and infrastructures from up close; to dwell in the familiar and really look at what is there to be seen.

The City Art Collaboratory challenges "the political economy of attention" that relegates and directs our habits of perception.[9] Such habits are not limited to artistic or scientific pursuits but structure

---

6 Timothy Morton, Ecology Without Nature (Cambridge, MA and London: Harvard University Press, 2007), 13.

7 Timothy Morton, *Ecology without Nature*, 13; and *Hyperobjects* (Minneapolis: University of Minnesota Press, 2013), 109.

8 Timothy Morton, *Ecology without Nature*, 27.

9 Sara Ahmed, Queer Phenomenology (Durham: Duke University Press, 2006), 32.

our everyday lives: what do we notice? What are we trained to miss? In her influential investigation of phenomenology, a branch of philosophy concerned with the body's capacities of perception, Sara Ahmed begins by observing the importance of where we are located: "Perception involves orientation; what is perceived depends on where we are located, which gives us a certain take on things."[10] How we find ourselves positioned, and perhaps more importantly, how we choose to position ourselves in the face of things, orients us. Where we choose to focus our attention, what we consider the foreground and what we relegate to the background, performs a certain magic: what can we see from where we stand? What fades into the background and becomes invisible?

Using Edmund Husserl's famous example of his writing desk, Ahmed shows the political economy of attention in action: Husserl's call to suspend the familiar, to look "as if" the table was a foreign object, as if we were entirely detached from its proportions, functions, and location erases the maintenance work that went into arranging the domestic space in this, and no other way: a door separates the philosopher from his wife, his children. Someone else ensures that his physical needs for sustenance are met, so he can think in the relative privacy of this study. Histories of labor along with the work of gendered bodies are relegated to the background, deemed irrelevant. Once such acts

10 Ibid., 27.

of relegation become perceptual habits, they effectively reveal how "some things are relegated to the background in order to sustain a certain direction," a certain set of priorities concerning what matters and what does not.[11]

If we leave Husserl's study and think about the city's infrastructures, the same economy of attention applies. Where and how we are located will orient us, which in turn will allow us to see certain things, identify particular problems, or not. Familiarity, far from being an obstacle to an outdated model of objectivity, is key in considering what we see: "Rather than the familiar being posited as that which must be suspended in order to see, we might consider what 'it' is that we 'overlook' when we reside in the familiar."[12] In contrast to artists forever on the go, the members of the City Art Collaboratory value the sensation of belonging to a place, of fostering a living connection that goes beyond the surface of things, a feeling for the place that is honed by familiarity and grows over time.

The City Art Collaboratory's insistence on the value of place finds support in Edward Casey's work. Bridging geography and philosophy, Casey considers place "the first of all things" and heartily disagrees with the argument that distance is a prerequisite for objective knowledge:

11 Ibid., 31.
12 Ibid., 34.

> There is no knowing or sensing a place except by being in that place, and to be in place is to be in a position to perceive it. Knowledge of place is not, then, subsequent to perception—as Kant dogmatically assumed—but is an ingredient in perception itself...genuinely local knowledge... is at one with lived experience if it is indeed true that this knowledge is of the localities in which the knowing subject lives. To live is to live locally, and to know is first of all to know the places one is in.[13]

Instead of deeming place a localist liability, Casey embraces emplacement as an epistemological necessity. The very notion of place is contingent on bodies: "bodies inhabit places; there are places only as lived by bodies."[14] Disavowing this connection impedes our ability to know, fully and intimately. What is at stake in place, then, is not the intellectual, Cartesian form of knowledge that retreats to the mind, but the immanent, experiential, embodied knowledge necessary for drawing on sensation and feeling as sources of relevant information about relationships with others and ecologies we may happen to be part of. Indeed, deliberately opting for displacement, applauding detachment from any putatively restrictive location, limits our ability to know. This is what we stand to lose

13 Edward Casey, *Getting Back Into Place* (Bloomington: Indiana University Press, 2009), 321.
14 Edward Casey, *Earth Mapping* (Minneapolis: University of Minnesota Press, 2005), xviii.

in the incessant mobility glorified as the epitome of flexibility in contemporary art and beyond.

Lived connections to place are not only relevant from an epistemological point of view. Our ability to know ourselves and others in relation to place may "recapture what the intellect has banned us from experiencing," writes Joanna Zylinska in *Minimal Ethics for the Anthropocene*.[15] Beyond that, though, having a relationship to place, an awareness of locatedness, is crucial for any sort of ethics: "Ethics is a mode of human locatedness in the world which involves a recognition of the processual and unstable nature not only of any such locatedness but also of the human that is thus located.... Ethics is not just about being-in but also about being with."[16] Being-in-place, grappling with the ever-changing intimacies of belonging, and its precarity, is prerequisite for an ethics, that is, for figuring out how to be and be with. Instead of applauding detachment, Zylinska advocates for a kind of thinking that originates "from amidst life itself:" that is the only way to ensure a place relatively free from hubris and normativity, entangled with complex particularities and inseparable from the universal, big-picture challenges we face.[17]

15 Joanna Zylinska, *Minimal Ethics for the Anthropocene* (Ann Arbor: University of Michigan Press, 2013), 53.
16 Ibid., 91.
17 Ibid., 21.

The City Art Collaboratory's commitment to place, then, does not limit and shackle creative and curious minds to a kind of doomed stasis. Instead, the Collaboratory's inquiries reveal that no place is ever stable but inevitably vested in processes: "places not only are, they happen."[18] Being located, aware of one's position, does not have to be detrimental to movement and process. But movement, in itself a mode of knowing, can happen on more humble scales and paces that exceed habitual human perception, such as the deep time of geology. Movement does not have to equal mobility along the pathways of prestige-pandering networks of cosmopolitanism. Knowing a place intimately and for a long time reveals place as a question. As Morton writes, "place as question is internal to the very question of self, of that which is located in place." More than a question of locality, place begs the question of who is located, when, where, why, and how: "The politics of place, then, is a struggle to achieve a state in which the question of place, the question that is place, can emerge as a question."[19]

Key for Morton's understanding of place as question is the idea that "place need not be a thing."[20] Indeed, according to his line of thinking, "place is radically indeterminate."[21] Far from concrete, place is an

18 Timothy Morton, *Ecology Without Nature*, 176.
19 Ibid., 177.
20 Ibid., 171.
21 Ibid., 174.

assemblage of tangible and intangible elements, processes, and structures that interact in often less than predictable ways. Place as assemblage acts on and affects us, in what Jane Bennett describes as emergent causality.[22] Understanding and studying the components of one such assemblage, one place, is what the City Art Collaboratory set out to do three years ago. The method of inquiry aims to be as spontaneous and adaptable as the subject of inquiry: a deviant interdisciplinarity, where scientific inquiry and creative license are free to mix and merge, cross-pollinate and inspire. What the City Art Collaboratory proposes, then, is a sense of place where such questions and modes of asking them can take shape.

22 Jane Bennett, *Vibrant Matter* (Durham: Duke University Press, 2010), 33.

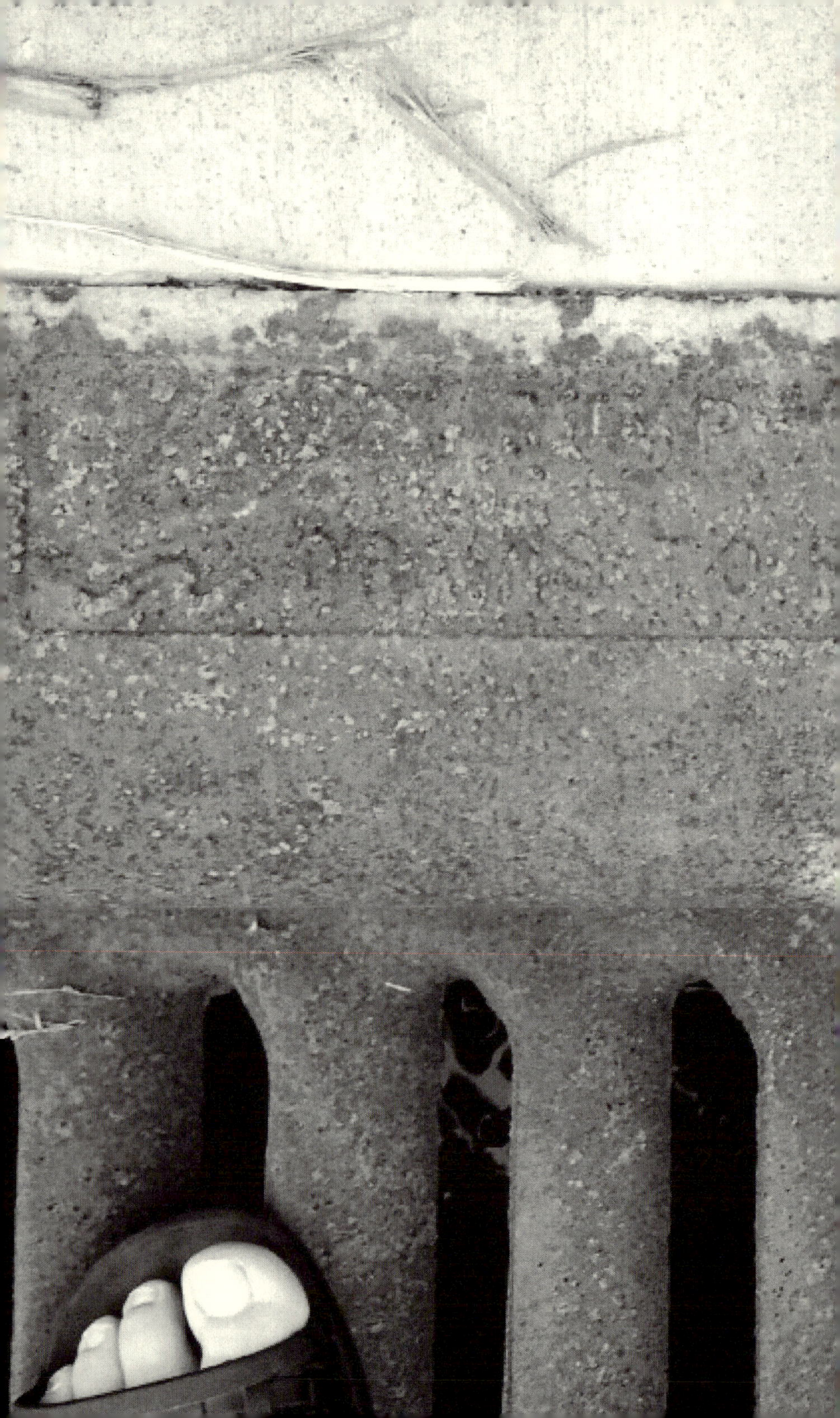

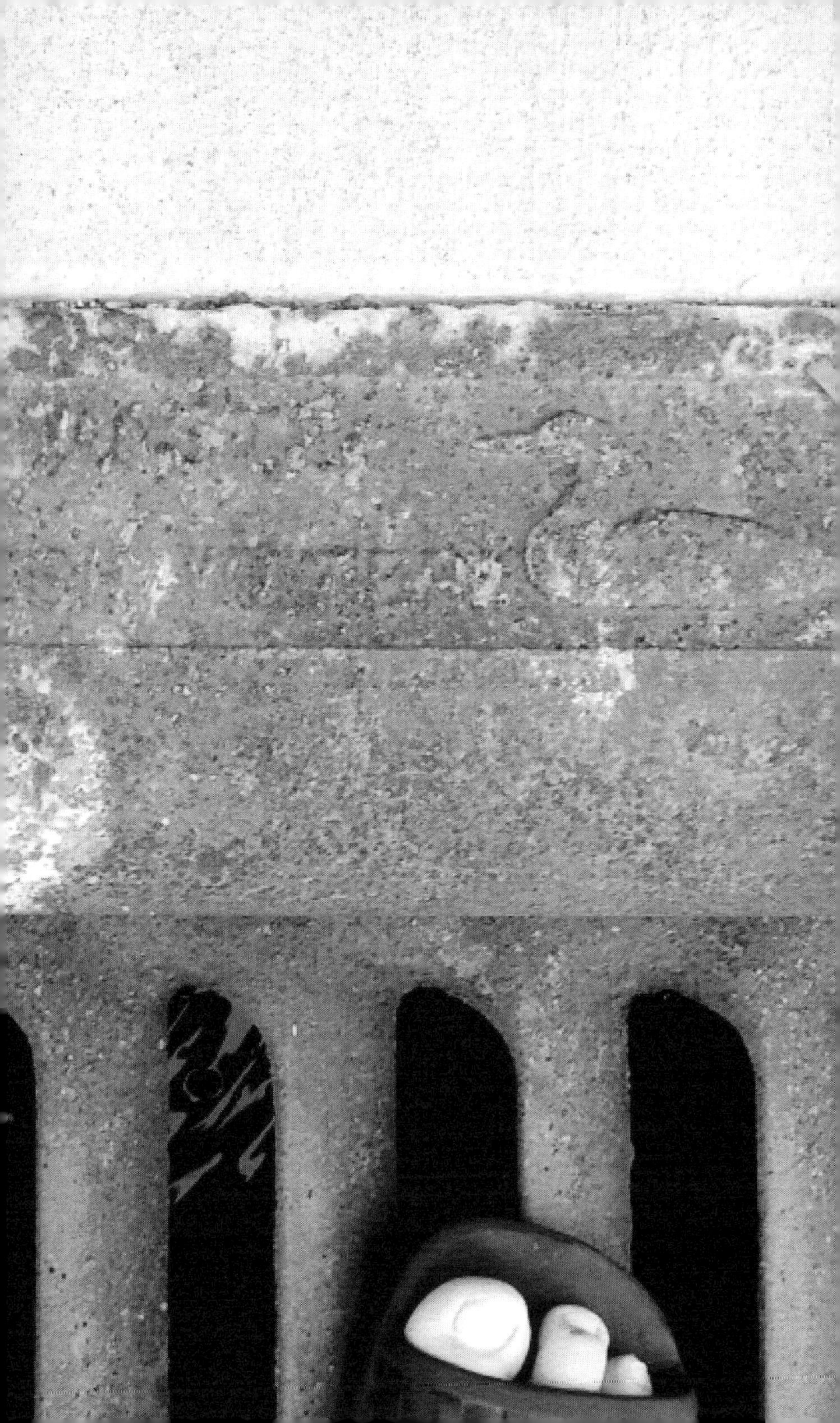

Preceding pages: A Collaboratory field trip along the Green Line, a light rail corridor connecting downtown Saint Paul with Minneapolis, led by Collaboratory Science Fellow Matt Kumka and his Barr Engineering colleagues in 2014. Photo: Andrea Steudel.

What I Learned from City Art Collaboratory

Patrick Nunnally

Patrick Nunnally is coordinator of the University of Minnesota's River Life program, which works to establish lasting relationships among those leading river sustainability efforts both on and off campus. Nunnally was an advisor to the City Art Collaboratory and a frequent participant in the group's activities.

I thought I knew a little bit about the Mississippi River, at least in the area around Minneapolis and Saint Paul. Years of going to meetings, building and hosting programs, and reading and writing, had given me a nodding familiarity with historical issues, questions, and problems with riverfront planning, as well as some of the organizational and institutional complexity of who does what on the river. I knew that there were vast realms I'd barely touched; many of the people I worked with had no real notion of the river itself other than that big wet area next to the park they wanted to develop or site on which to build. Likewise, I'm no scientist and barely understand the difference between hydrology and fluvial geomorphology.

Often I've found myself the only person in a room who was more than minimally conversant in multiple river conversations: the only historian in a room full of bike planners, or the only public engagement person in a room full of scholars.

In short, after working on river stuff for about fifteen years, I thought I had a decent conceptual map of the field, who is doing what, and how those discussions were coming along.

Then I started hanging out with the City Art Collaboratory and realized the extensive limits to my understanding. To put the matter over-simply, I knew a bunch of static, individualized things, small conversations that were self-contained, grounded in shared starting points and understandings and urgencies, with generally agreed-upon goals.

What I needed to know (following the "what I know" / "what I want to know" model of Give & Take, a program by Works Progress Studio) was how to put these pieces in motion, connect them, fill the spaces between diverse conversations, and see things I had not known to be looking for—the "unknown unknowns," if you will.

Let me give you some examples:

On its first Field Trip, the City Art Collaboratory group assembled on a weekend afternoon at Lock and Dam No. 1, otherwise known as Ford Dam. We wandered around the facility, talked to the guys working there from the Army Corps of Engineers, and conversed among ourselves. It was a beautiful afternoon to be out and about, and the group adjourned to a nearby city park community room to continue the discussion.

I had been reading materials and attending "Asian Carp Task Force" meetings for some months at this point, addressing the issue of how to manage and mitigate the invasive species. I had actually done some research on the lock facility during a summer job with the Corps back in the 20th century. So I thought I was all set with some "expert knowledge" that would be useful to the rest of the folks who were coming there for the first time.

I was wrong.

No one really wanted to know the details about the carp, which, after all, are pretty far downstream. Sure, there was interest in a general sense about how people were variously reacting to a perceived threat, but no one really wanted to listen to a detailed explanation of what agency has responsibility for what parts of the work. And no one at all wanted to know the history of the facility and how that history fit into the efforts to remake the Upper Mississippi into a better conduit for shipping. It's as if the mythical, badly-taught high school history class—with names and dates disconnected from any life now—was continuing to grip us.

Some months later, I joined another Collaboratory Field Trip, this time to Bruce Vento Nature Sanctuary. The Vento Sanctuary encompasses part of the area known to Dakota people as Wakpá Tháŋka, roughly "house of spirits," which is in the river floodplain below the Mounds Park site managed by the City of Saint Paul. This is a remarkable, near-magical place, one that inspires more listening and looking than talking. Our group was quite diverse and included Dan McGuiness, a long-time Mississippi River leader who was serving at the time as the Executive Director of the Lower Phalen Creek Project, a community group that coordinates restoration activities at the site.

During our visits there, I heard more than I can say. What I learned I haven't yet processed enough to become a written part of this story. The point is, though, that the artists in the Collaboratory were part of the expansiveness of the discussions, which kept this from being a meeting reduced to "objectives" and "so therefore" next steps.

In April 2014, my organization, River Life, held an event called "The Irony of Carp." While this wasn't really a Collaboratory event, it did prominently feature a Collaboratory Art Fellow, Molly Balcom Raleigh.

Molly was part of a panel that responded to the main speaker, Brian Ickes from the U.S. Geological Survey.

Ickes is an internationally-recognized scientist specializing in invasive carp, and his presentation focused on some of the ironies of the broadly based effort to stop the carp migration up the Mississippi River. To take just one example, Chinese fishery experts cannot sustain a population of these fish in the Yangtze River, where they originate, because of pollution and heavy shipping traffic that disturbs the waters. The irony is that the most robust population of the fish is in the Illinois River in the United States, where agencies are spending millions of dollars trying to contain and eradicate them.

Molly's take on the subject was distinctive, and here we see one of the things artists can bring to conversations they take part in. She had read some of the "stop carp" materials and seen some videos from the campaign online. She was struck by the language and tone that she saw, and pursuing this inquiry further, realized the striking similarities between "stop carp" rhetoric and the xenophobic rhetoric about human "illegal alien" immigration to the United States.

Of course, this event, which took place in April 2014, predates the presence of Donald Trump as a presidential candidate in our public awareness. What we now understand as a pervasive ugliness in public discourse on one subject was recognized some years ago as an unintended consequence of much more targeted discussions on what was thought to be a completely different subject.

What do I think I know now:

Here I am, after several trips with the City Art Collaboratory folks. I had a great time, met interesting and stimulating people, and learned a lot. Namely:

- Programs are better if we have a general orientation but leave space for new perspectives;

- Having people in the room who are coming from different spaces but are better listeners than tellers really helps;

- We probably don't really know what the upcoming problems are yet, much less how to "solve" them;

- What we know isn't as important as what we haven't learned yet;

- Art isn't about making a thing, it's about making a thing; obvious to some, news to many.

We need artists in all of these public conversations about the Mississippi, about place, community, water, and identity, for the same reasons we need indigenous people and their perspectives: they are important in their own right, and they push our boundaries in ways we didn't know we needed to be pushed. The conversations about our collective future here are only as rich as the breadth of knowledge of the place (from those who were here first), and as wide as our imagination allows them to be. Artists disturb what we think we know, and by doing so, help us know better.

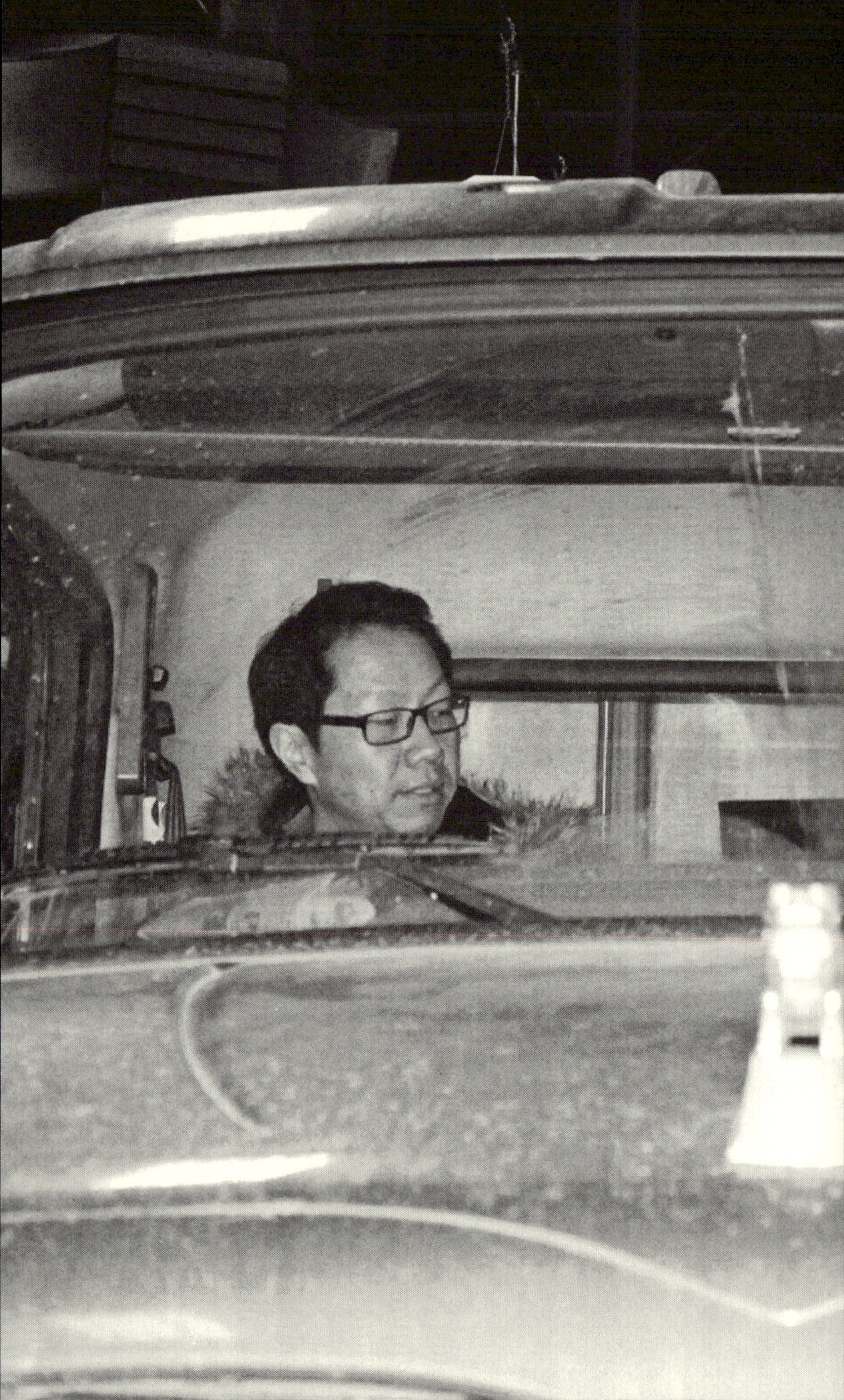

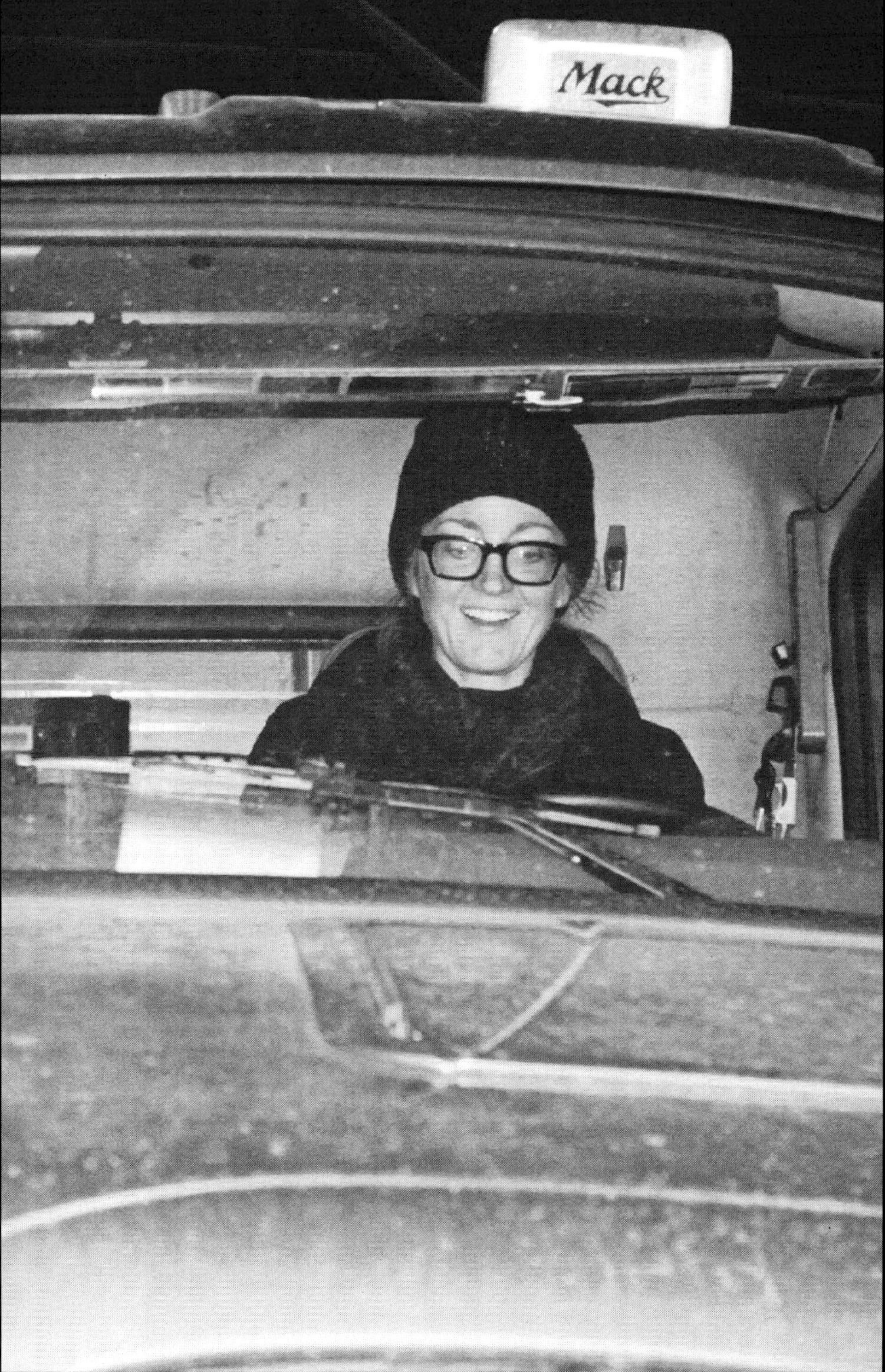
Mack

Preceding pages: Collaboratory Curator Marcus Young and Art Fellow Emily Stover on a field trip to the City of Saint Paul's Street Maintenance Division in 2013. Photo: Andrea Steudel.

Four Years of City Art Collaboratory

City Art Collaboratory began without a set schedule of activities or publicly oriented projects. As organizers and curators, we were guided by goals of process more than content. We aimed to create an open-ended and encouraging space for artists and scientists to better know themselves and one another, to ask and pursue important questions, and to incubate collaborations in places and systems that touch the Mississippi River.

At times we struggled to communicate what exactly a City Art Collaboratory is and does. Is it a fellowship program? Yes, we did cultivate a sense of belonging and fellowship among a group of people interested in similar things. Is it a think tank? Yes, we encouraged Collaboratory Fellows to think together, and to apply their creative ideas to real situations in and around Saint Paul. We also went outdoors, experiencing our immediate environment on foot and in canoe. Occasionally, we made things together.

Over time, some of these ideas and relationships developed into new or expanded collaborations. We encouraged this kind of collaborative development by creating opportunities for public presentation, peer review, and resource sharing among Collaboratory Art and Science Fellows, as well as other groups.

In retrospect, City Art Collaboratory became a means of slowing down, reorienting and catalyzing art and science in an emerging community of practice. It may seem counterintuitive, but to develop new ways of thinking, making, and living that address complex and systemic ecological problems, we need to meander awhile. To quote Dakota artist Mona Smith, whose indigenous thinking about place has inspired many in and around Saint Paul: "Know who you are, know where you are," and know who you are with.[1]

What did this process look like for City Art Collaboratory? The following pages document some of our group activities, which included informal gatherings around meals and in the homes and workspaces of participants; field trips and design workshops within city systems; public events showcasing works in progress; and support for open-ended research and creative experimentation. Throughout Collaboratory's first four years, we did our best to listen, adapting the trajectory of the program so that it could respond to the emerging questions and curiosity of those involved.

---

1 Jan Willms, "Memory Maps" in *Minnesota Women's Press* Online, September 6, 2016, http://www.womenspress.com/main.asp?SectionID=1&SubSectionID=20&ArticleID=4015.

First Gathering
9/8/12 - Saint Paul, MN

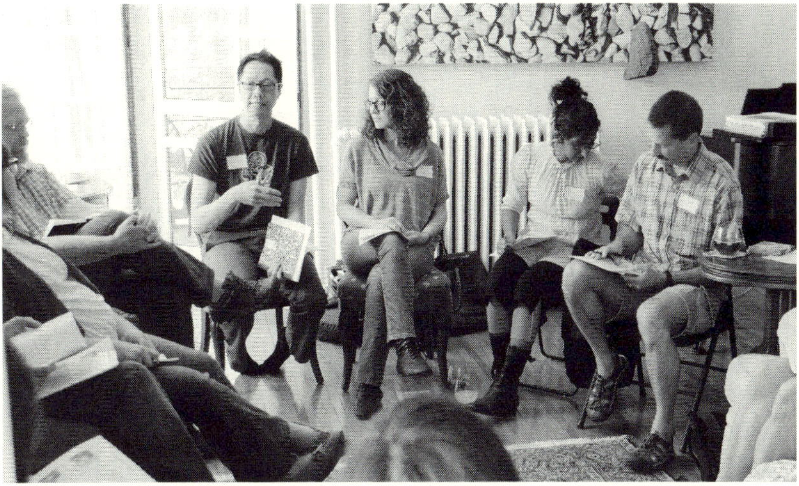

At this first gathering of the City Art Collaboratory, we shared a meal and introduced ourselves. In one-on-one conversation and in small groups, Collaboratory Fellows discussed their initial questions and goals for participating in this newly launched program, coming up with a list of field trip sites they wanted to visit, as well as themes for the year. We used this input to develop field trips and related activities.

Photo: Kent Larson

"What new insights and connections are possible among a group like this? How do we combine intuition and analysis to create something new? How can public art impact the sustainability of buildings? How can we bridge the perceived or real gaps between science and art? What are the historical connections between natural resources, national identity, and power? When we're talking about ecology and things like restoration of ecosystems, who and what belongs? When we look at the river, what is invisible to us? How can we foster creative misunderstandings?"

From a list of questions collected at our first gathering.

Lock and Dam No. 1 (Ford Dam)

11/18/12 - Saint Paul, MN

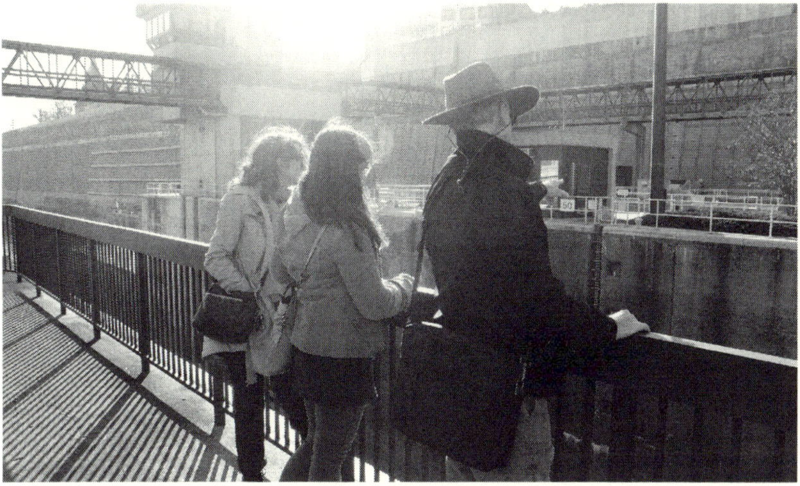

For our first field trip, we brought Collaboratory Fellows to Lock and Dam No. 1 (also known as Ford Dam) on the Mississippi River. They were joined by research scientist Peter Sorensen from the University of Minnesota and John Anfinson of the National Park Service. Along with employees of the Army Corps of Engineers, these outside experts introduced Collaboratory Fellows to one of the issues defining the future of ecology and economy on the Mississippi River: invasive species, including several species of carp.

Photo: Andrea Steudel

"One moment in particular that I remember well is a group conversation around aquatic invasive species, which highlighted for me the notion that artists can bring an entirely different perspective to a situation, and critically challenge current ideology that primarily stems from a scientific perspective. Plus, it ended with a carp dinner!"

Leslie Yetka, Collaboratory Science Fellow

Metropolitan Wastewater Treatment Plant

12/13/13 - Saint Paul, MN

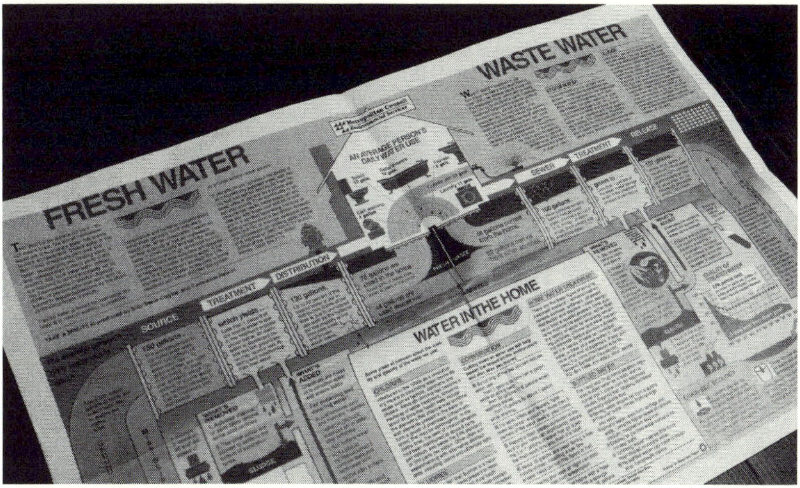

Our second field trip brought the Collaboratory to the Metropolitan Wastewater Treatment Plant, the largest wastewater facility in Minnesota, which serves 1.8 million people and discharges into the Mississippi River. Fellows received an extensive tour and Q&A with Larry Rogacki (Metropolitan Council Environmental Services). Though photos were not allowed on site, many brought their sketchbooks and documented the trip through illustration.

Photo: Shanai Matteson

Collaboratory Social

01/19/13 - Saint Paul, MN

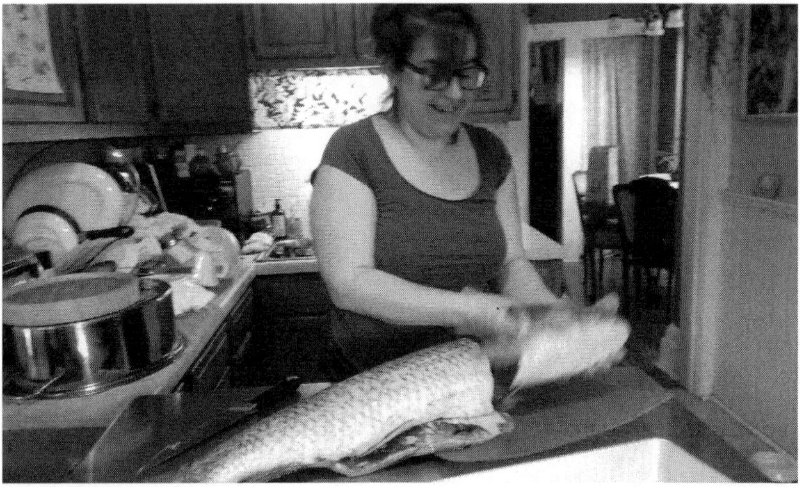

To encourage social connections among members of the Collaboratory and Public Art Saint Paul staff and supporters, we held a gathering at the home of Public Art Saint Paul Board Member Bob Muchewske and his wife Leaetta Hough. At this gathering, Collaboratory Art Fellows Molly Balcom Raleigh and Janaki Ranpura sourced and prepared a meal of invasive carp, recalling the issues discussed on our first Collaboratory field trip, some of the proposed solutions, and new ways of imagining the problem.

Photo: Janaki Ranpura

City of Saint Paul Street Maintenance

03/02/13 - Saint Paul, MN

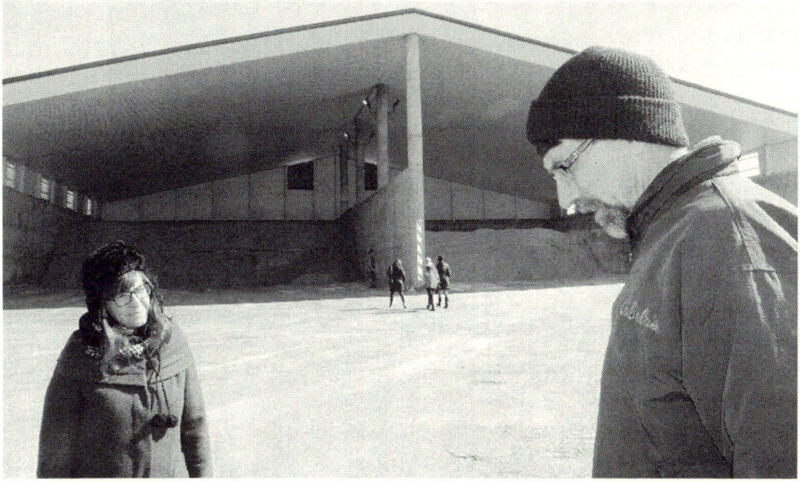

The Collaboratory cohort met for a field trip to the City of Saint Paul's Street Maintenance Division, which is responsible for snow plowing and salting. In addition to learning about the inner workings of a vital public works operation, Collaboratory Fellows heard about how the city communicates during weather emergencies, how climate change is already having an impact on city services, and the ways salt and sand compromise water quality in lakes and streams.

Photo: Andrea Steudel

"After Collaboratory, I think more about the people who work to keep a city running, and about all of the decisions, big and small, that have shaped Saint Paul. It breeds a different kind of appreciation when you see these people and systems working in real time."

Emily Stover, Collaboratory Artist Fellow

Lowertown Walking Tour
04/04/13 - Saint Paul, MN

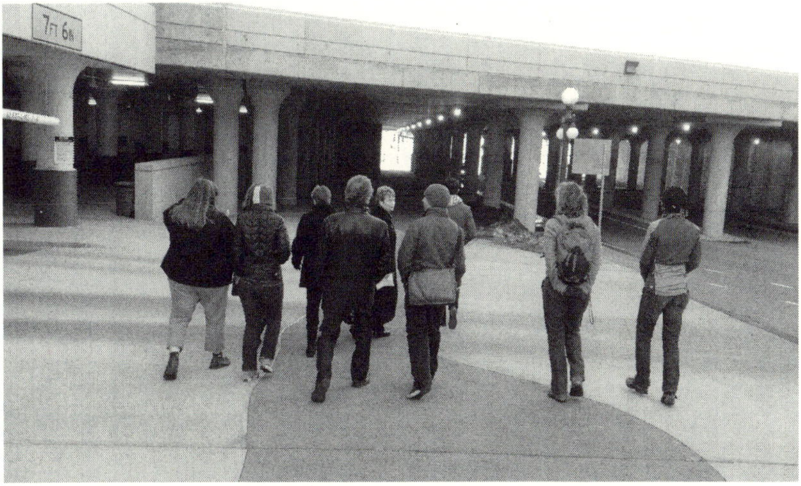

Collaboratory Fellows took a walking tour of Lowertown, a neighborhood where access to the Mississippi River is vital to the future. This tour was led by Saint Paul City Planner Lucy Thompson, Environmental Policy Director Anne Hunt, and Tim Griffin of the Saint Paul Riverfront Corporation. Afterward, the group convened at the Minnesota Museum of American Art for dinner and a workshop to generate art and engagement ideas, as well as new research questions.

Photo: Andrea Steudel

"I think it is critically important to bring a variety of ways of seeing to projects and planning initiatives that involve hard infrastructure and critical services, and impact our landscapes, but that ultimately must involve and depend upon people."

Amanda Lovelee, Collaboratory Art Fellow

Mississippi River by Canoe

08/05/13 - Saint Paul, MN

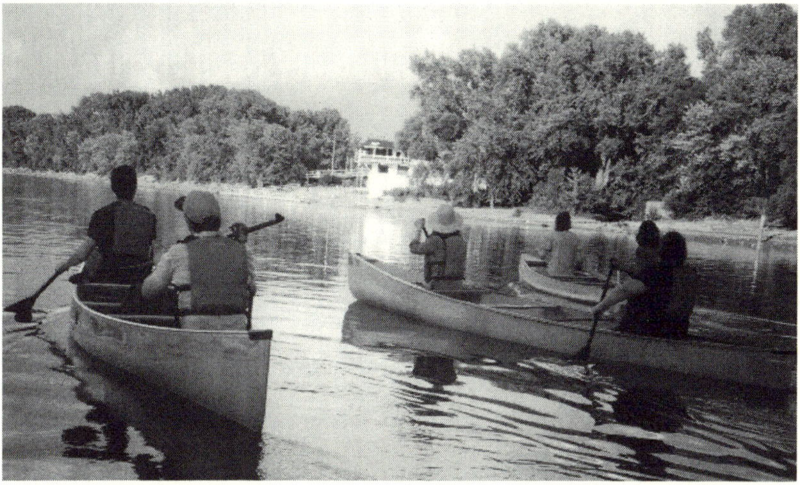

Members of the Collaboratory stated a goal of getting closer to the Mississippi River, sharing outdoor experiences and deepening their relationship to place. On this field trip, we canoed on the river from Hidden Falls Park to downtown Saint Paul, observing how close contact with water and the collaboration required to pilot a canoe both change perception of place.

Photo: Shanai Matteson

"We might know one lens, but through seeing them all at different times, the scale of the river becomes more clear. Beyond the size and length of the largest river in the U.S., its ecology is larger than we can capture. The ecology of the Mississippi touches the history of the Dakota people, is the reason western people came together here, it touches the history of the commerce of the cities, creates jobs, and the changes made to the river allow for ecological disasters to occur. I would not have understood all of that without the Collaboratory, and I have since applied that kind of thinking to other places, systems, and ecologies."

Katie Hargrave, Collaboratory Art Fellow

2013 International Low Impact Development (LID) Symposium

08/23/13 - Saint Paul, MN

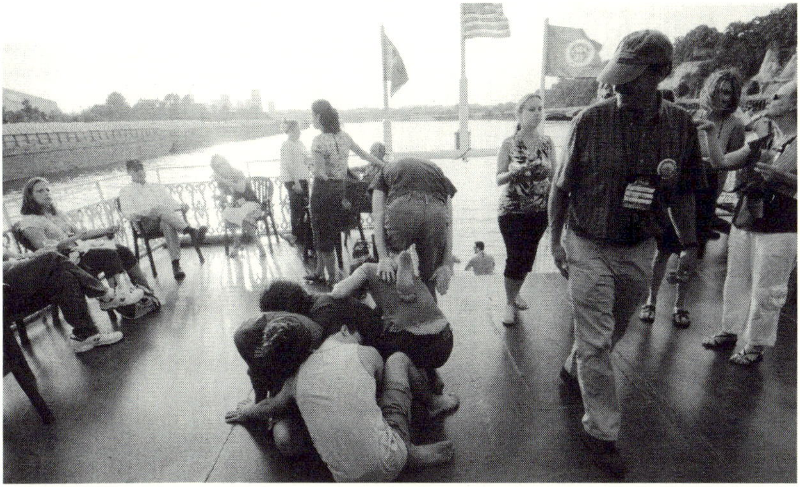

Environmental artist Christine Baeumler and Public Art Saint Paul both supported Collaboratory Art and Science Fellows attending the LID Symposium in Saint Paul. Public Art Saint Paul also sponsored a boat ride on the Mississippi River for Collaboratory Fellows and other LID Symposium attendees. Artists from the Collaboratory planned creative activities aboard the boat to engage and connect with symposium attendees.

Photo: Sean Smuda

Collaboratory Conversation Platform

08/21/13 - Saint Paul, MN

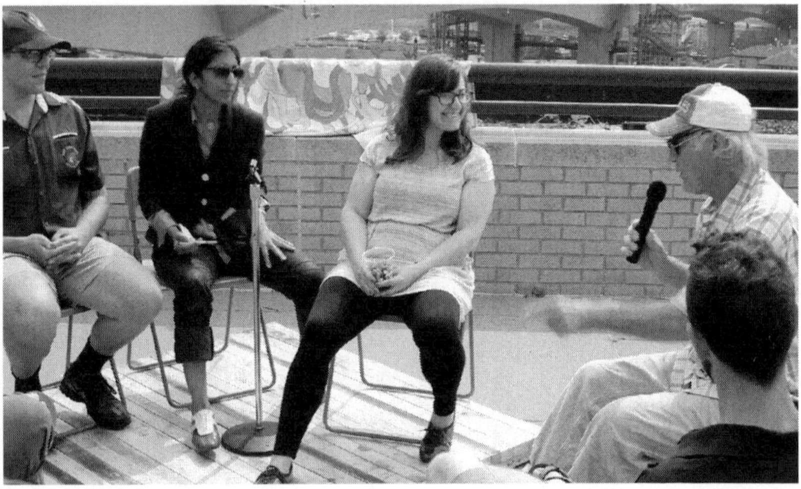

Collaboratory Fellows built a simple wooden platform and used it to convene public conversations during lunch breaks in downtown Saint Paul, as part of their participation in the LID Symposium. Artist Janaki Ranpura led these public conversations, which featured international experts from the LID Symposium, including artist Buster Simpson, pictured here.

Photo: Shanai Matteson

"Our public conversation had an interesting, idea-tossing delightfulness that we might normally keep behind closed doors, in a private Collaboratory meeting. It was great to have it out-of-doors, for anyone who got interested enough to listen... The conversation's participants felt, I think, a level of focus in the talking that was associated with it being performative, in public space, potentially for a larger group of unknown people."

Janaki Ranpura, Collaboratory Art Fellow

Ordway Research Station
10/10/13 - Inver Grove Heights, MN

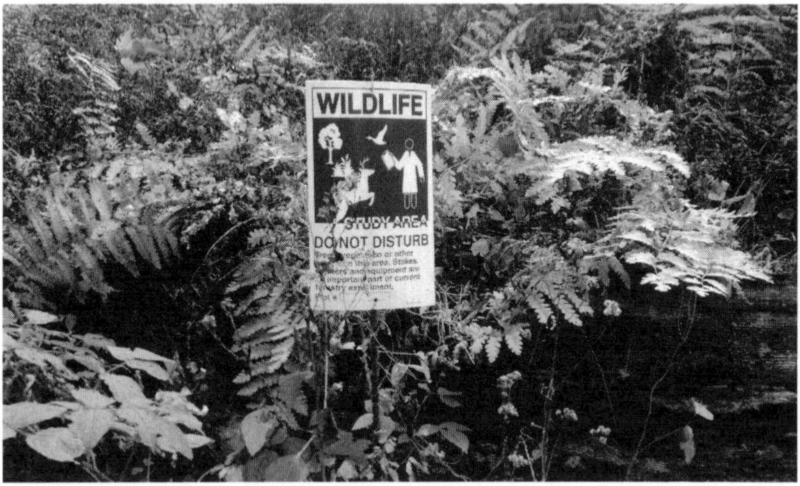

One of the priorities identified by the Collaboratory cohort was learning more about one another. This field trip brought the group to the Katharine Ordway Natural History Study Area, which is managed by Collaboratory Science Fellow Jerald Dosch. The group was able to hike trails and get out onto boats within this natural history area, learning more about ecological research and science education.

Photo: Shanai Matteson

Public Gathering at City House

11/01/13 - Saint Paul, MN

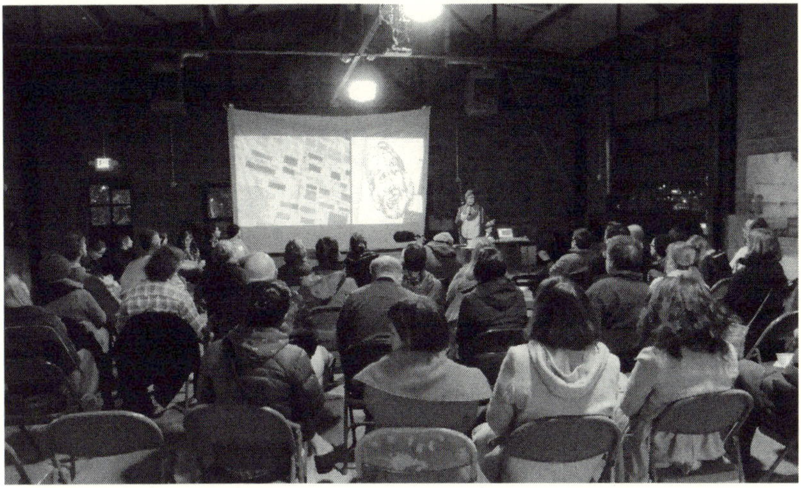

At the close of 2013, Collaboratory Art and Science Fellows shared their experiences, questions, and ideas at a "pitch night" and public gathering. This event, held inside a vacant grain terminal on the Mississippi River, introduced the Collaboratory cohort to a public audience and served as a capstone for year one of the program.

Photo: Andrea Steudel

"The Collaboratory has little to do with measurable outcomes or economic development. It's an investment in artful city-making with an eye on the long view: it's about building relationships, and trust, across the boundaries of disciplines, academic paradigms and modes of working. And it's about the promise of serendipity that might spring from the creative friction generated by such thoughtful cross-sector partnerships."

Susannah Schouweiler, "Saint Paul's new art-science 'Collaboratory' holds their first public gathering," Knight Arts Blog

Ripple Impact Workshop

12/16/13 - Saint Paul, MN

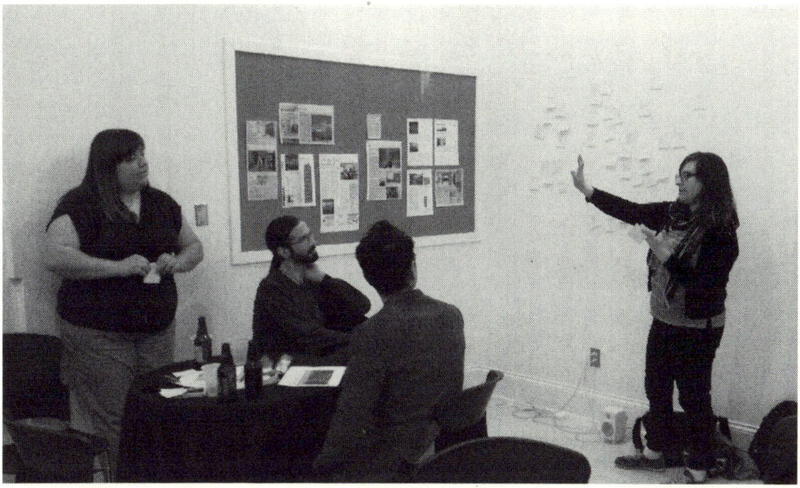

Following year one of the Collaboratory, the group participated in a sense-making and impact mapping process designed by Public Art Saint Paul Development Director Eric Asboe and Collaboratory Director Shanai Matteson, based loosely on Ripple Impact Mapping techniques, which are explained in more detail on page 133 of this book.

Photo: Shanai Matteson

"Many of these impacts are intangible and manifest in the language I use to describe my intellectual interests and how I talk to students about their own work. It has influenced how I mentor and who I mentor, the breadth of ideas that I consider part of my scholarship. My conversations with Collaboratory Art Fellows Aaron Dysart and Olive Bieringa, in particular, have helped me see how artists can provide a sensory experience of something that I would not have thought possible without special scientific instruments to extend the senses. For me, this was a profound insight that has set me on a new path of discovery and has inspired me to explore my own creative aspirations with a new passion."

John Schade, Collaboratory Science Fellow

Retreat with Second Collaboratory Cohort

02/01/14 - Hastings, MN

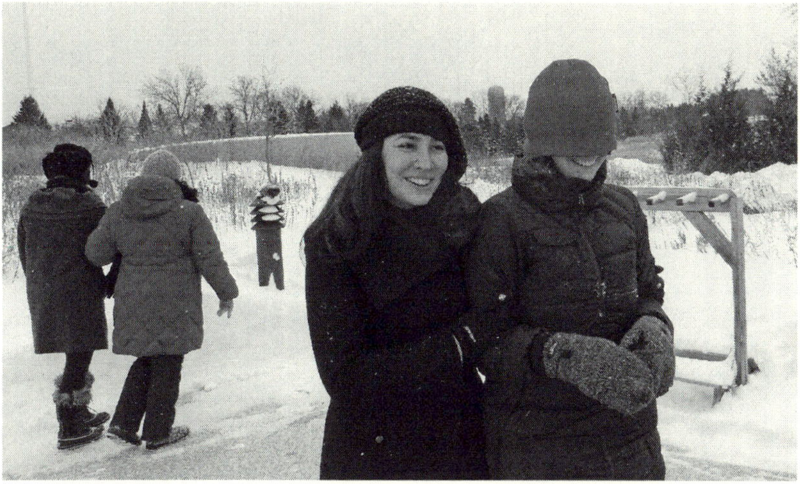

The group kicked off the second year of City Art Collaboratory with a retreat to Schaar's Bluff Gathering Center in Hastings, Minnesota. Each Fellow gave a flash presentation about their work and the questions they were bringing to the group, using images or props as appropriate. The group also spent one-on-one time outdoors, prepared a meal together, and participated in an activity designed to explore the varied language we use within our respective disciplines to describe place and process.

Photo: Andrea Steudel

Collaboratory Socials

Throughout 2014 - Amsterdam Bar and Hall, Saint Paul, MN

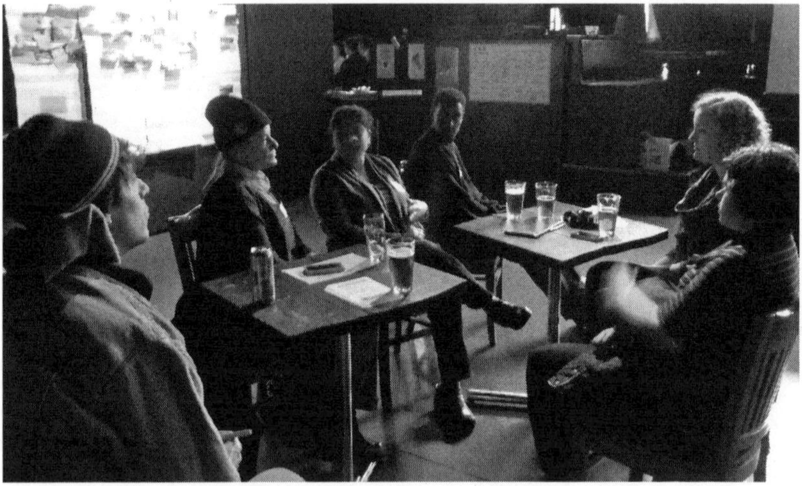

Each month throughout 2014, a different artist or scientist from within the Collaboratory cohort planned or presented a social gathering at a downtown Saint Paul bar and restaurant. This social space provided a platform for public sharing of works-in-progress, and a way to expand the conversation that began within Collaboratory to others outside the group. An example of one of the ideas shared through Collaboratory Social, a presentation from Collaboratory Art Fellow Janaki Ranpura, can be read on page 209.

Photo: Andrea Steudel

"I was drawn to this because I am deeply interested in how scholars can be more engaged in public art. I find that in environmental studies we need to better consider how to bring the broadest publics in to help define our research questions, help us interpret the trends we see and push out our findings so that we can impact people's day-to-day lives. Publishing only gets us so far in changing the world."

Roopali Phadke, Collaboratory Science Fellow

District Energy

03/01/14 & 3/29/14 - Saint Paul, MN

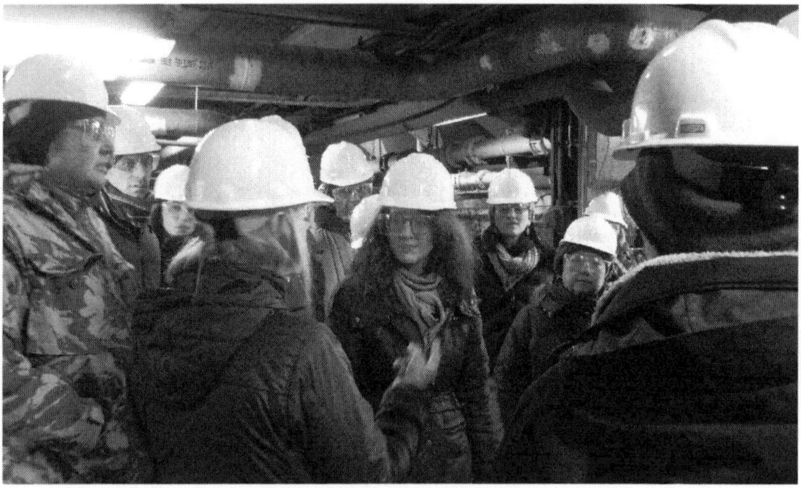

In our second year, we used feedback from participants to design activities of different types. We chose to alternate between field trips to specific sites or organizations, and workshops with leaders and community members involved in those sites or organizations. Our first field trip of 2014 brought the Collaboratory to a combined heating, cooling, and electricity plant on Saint Paul's riverfront. Among other new ideas, this tour inspired the development of *The Plume Project*, profiled on page 147. It also led to some difficult conversations among the group about the importance of making sure that collaborative ideas are sparked, developed, and shared in open but equitable ways.

Photo: Kirsten Valentine Cadieux

"Our field trips were informative and inspiring and left me thinking about the relationship between science and art. I realized that one major thing we must do as artists is to educate the public about what we do. It seems many think that artists just make drawings and paintings, or even cartoons—yes, we do that sometimes, but we do so much more as well! Likewise, I am sure I have misconceptions about what scientists do, and the Collaboratory has been a way to expand understanding."

Aaron Marx, Collaboratory Art Fellow

Bruce Vento Nature Sanctuary

04/26/14 & 05/31/14 - Saint Paul, MN

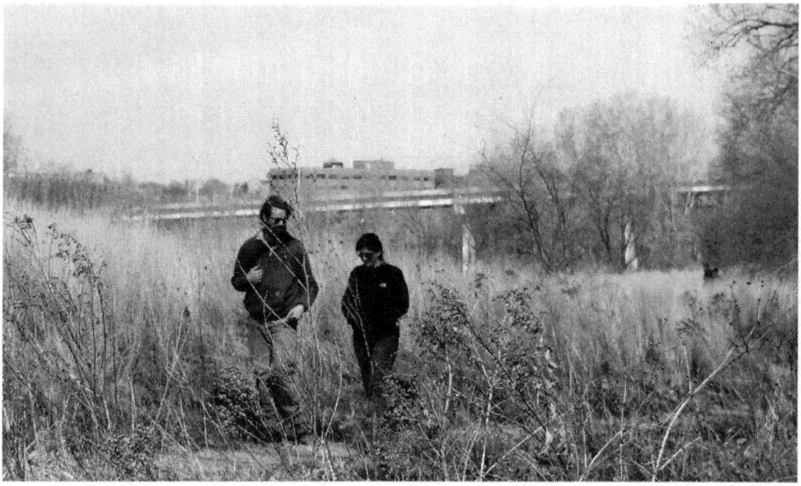

Collaboratory Fellows joined artists, activists, and community residents at Bruce Vento Nature Sanctuary, also known as Wakpá Tháŋka (roughly "house of spirits") to Dakota people. This area has also been a rail yard, and effectively, a trash dump. In recent years, community residents and groups have come together to restore the site, and are in the process of charting a future that includes a new interpretive center. This field trip introduced Fellows to this process and opportunities to get involved.

Photo: Andrea Steudel

BodyCartography Activity at Bruce Vento
06/28/14 - Saint Paul, MN

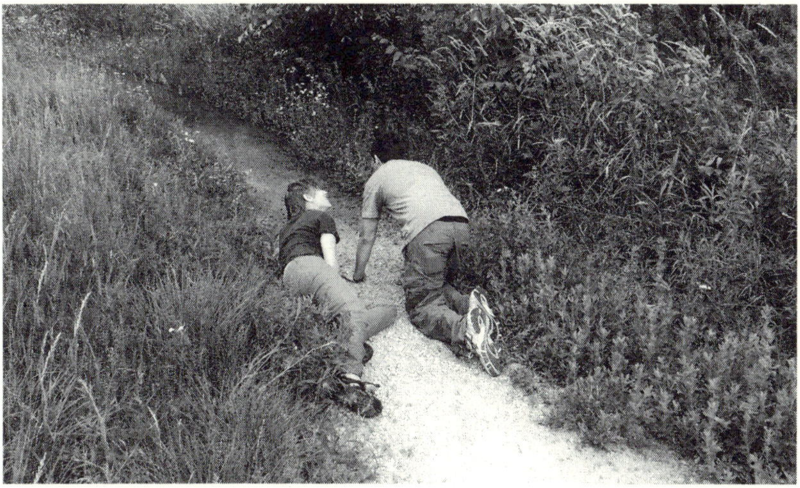

In the second round of Collaboratory field trips, participants asked to gather for more frequent creative activities. Dance artist and Collaboratory curator Olive Bieringa led a movement exercises to build empathy and relationships at this important field trip site.

Photo: Andrea Steudel

Green Line Tour

08/23/14 - Minneapolis & Saint Paul, MN

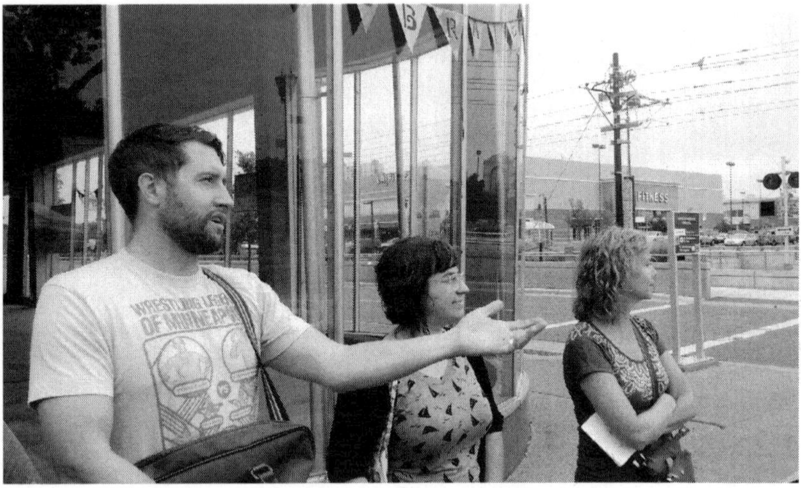

Collaboratory Science Fellow Matt Kumka led a tour of the Green Line, a light rail corridor connecting downtown Saint Paul with Minneapolis. This tour focused on environmental design and engineering projects along the line that were built to improve urban habitat and water quality. This tour also featured the work of artist and botanist Sarah Nassif, whose project *The Other Green Line* explores the natural history of the same corridor.

Photo: Andrea Steudel

Coldwater Spring

10/25/14 - Saint Paul, MN

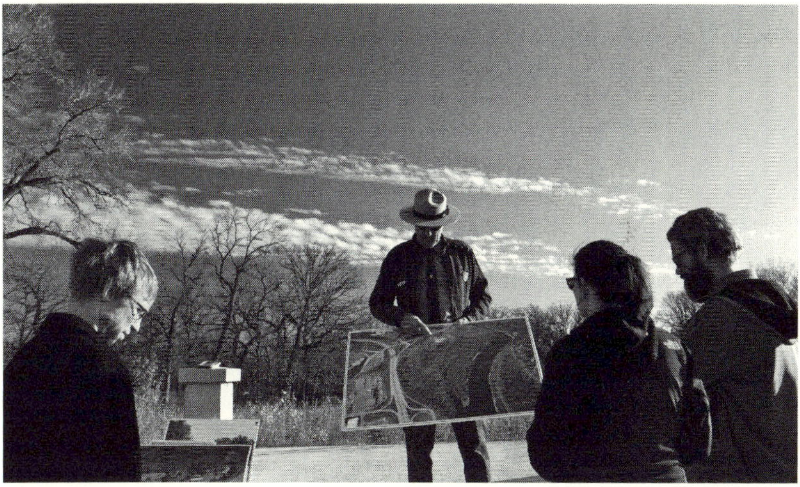

On the final Collaboratory field trip of 2014, Fellows toured Coldwater Spring, an area along the river in Minneapolis that has been transformed many times over the years. A longer entry about this particular field trip by Collaboratory Director Shanai Matteson can be read on page 113, and a critical response by Dan S. Wang can be found on page 121.

Photo: Andrea Steudel

Workshop: Making a Manifesto

02/08 - 02/14/15, Saint Paul, MN

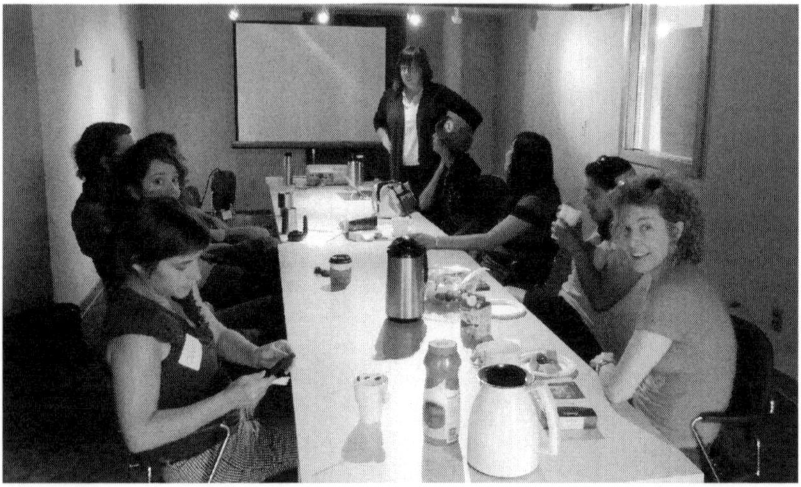

Collaboratory Fellows convened to close this phase of the program, working with writer and educator Christina Schmid to author a group artist statement. This "retroactive manifesto" can be read on page 39, as well as Schmid's critical take on the Collaboratory on page 45.

Photo: Andrea Steudel

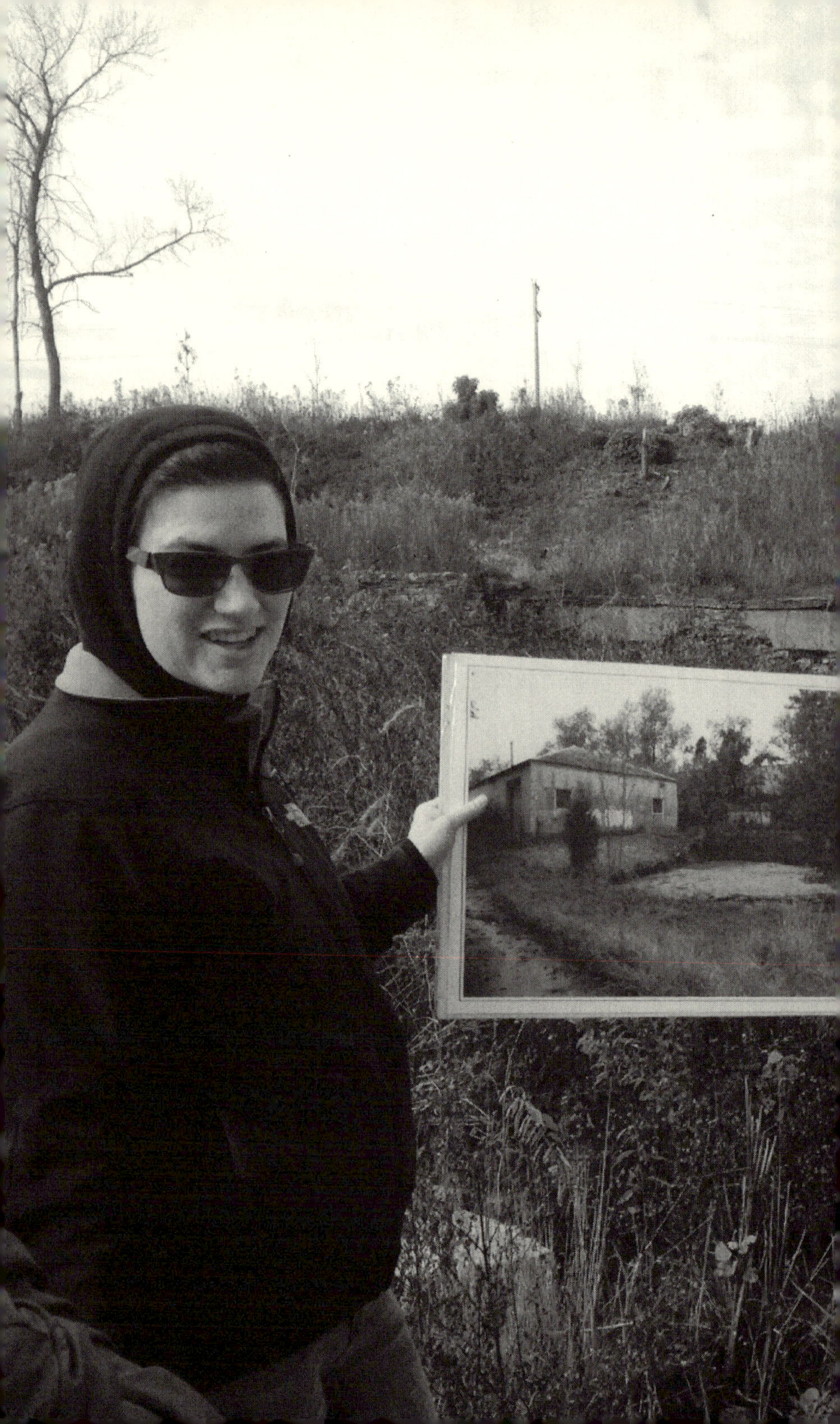

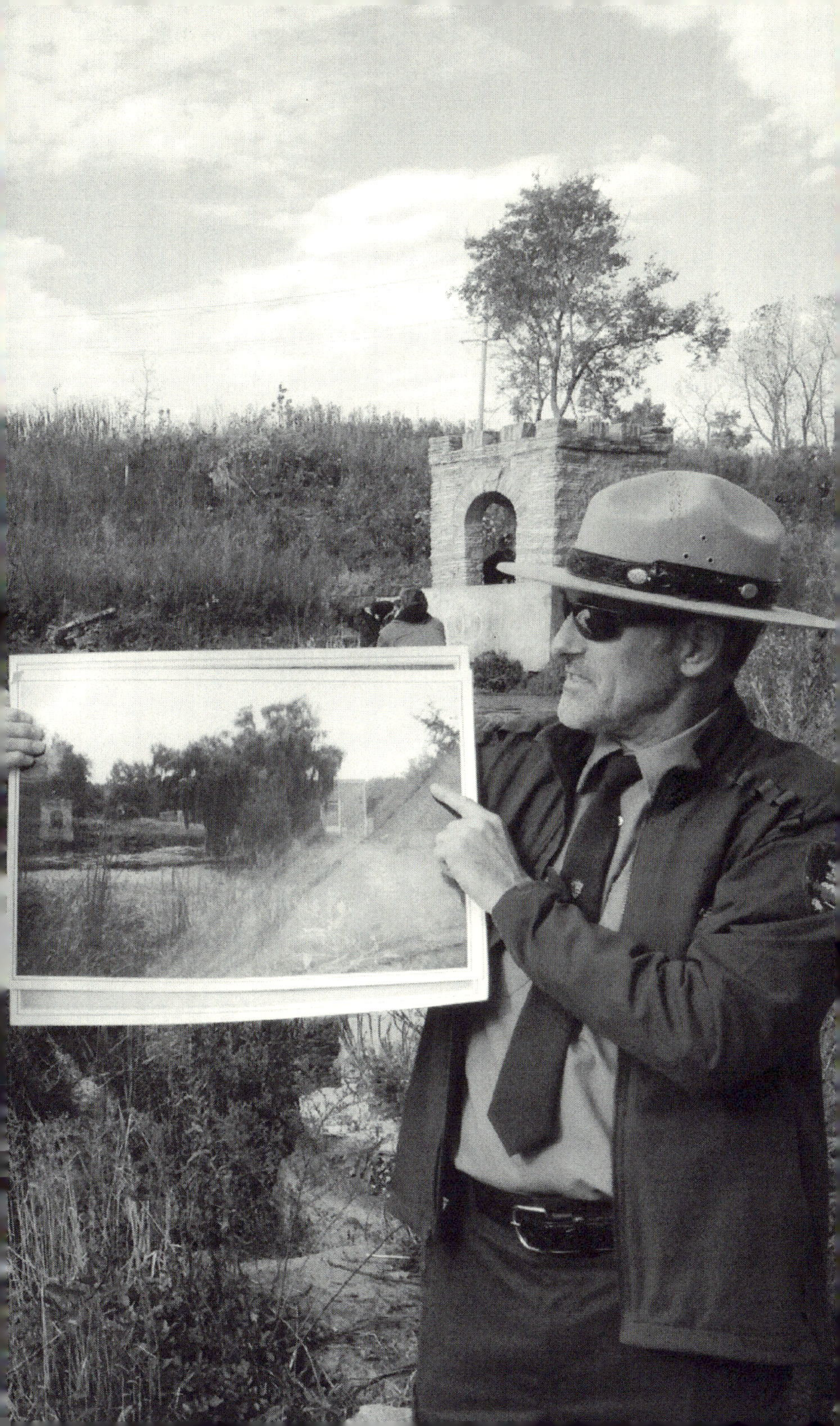

Preceding pages: Collaboratory Science Fellow Kate Brauman and National Park Service Ranger Alan Robbins-Fenger on a field trip to Coldwater Spring in 2014. Photo: Shanai Matteson.

Field Trip Recap: Coldwater Spring

Shanai Matteson

Reflections from each City Art Collaboratory field trip were shared through a blog created for the project. This post documents a visit to Coldwater Spring, included here as it relates to the subsequent essay by Dan S. Wang.

For its final field trip of 2014, the City Art Collaboratory cohort met at Coldwater Spring, a parcel of land granted to the National Park Service by the Federal Government in 2010. Since that time, Park Service staff and partners have been working on ecological restoration of the site, taking down old buildings, removing debris and contaminated soil, mitigating invasive species, planting trees and shrubs, and seeding acres of prairie and wetland. Pavement has been replaced with walking trails, and it remains a special place to spend time.

Alan Robbins-Fenger from the National Park Service met our group and gave us a tour. He shared information about the restoration efforts of the Park Service, as well as a brief history of the site, which is extremely complex. He also answered our questions about the process of acquiring and maintaining such a significant parcel of land, including the public engagement protocols of the Park Service and similar Federal Agencies.

Why is this place significant? Layers of history, ecology, and meaning converge at Coldwater Spring. This land sits near the confluence of the Minnesota and Mississippi Rivers, a place that is central to the spirituality and history of Dakota people. The spring itself has long been a source of fresh drinking water for people living in the area, and some believe it is sacred and has healing properties. It was also an important water source for the military outpost established at Fort Snelling during the U.S. Army's westward expansion, giving the site added significance in terms of water transport and storage technology, and for its role in the urban development of the Twin Cities.

More recently, the site was home to the Bureau of Mines, a research and development lab affiliated with the University of Minnesota. The buildings that once stood on the property included a core sample library and headquarters for the Bureau, which developed safety equipment and other technologies used worldwide in mining, including methods of pelletizing taconite to make it safer to transport over land and water.

During the 1990s, when plans for the rerouting of Highway 55 were being developed, Coldwater Spring became the site of the longest-running protest in Minnesota history. Activists concerned about the natural and cultural significance of the property protested the highway reconstruction, occupying the area and building an encampment in an effort to halt development. In 2010, after some vigorous public debate, the Park Service was granted the property, which had been neglected for many years. It became the first "on land" park property in the Mississippi National River and Recreation Area, a National Park that spans the Twin Cities along the Mississippi River. Previous to this, the Park had contained only river island properties, some of which disappeared when inundated during periods of high water.

We spent much of our time talking with Alan about the complicated and hybrid nature of the site in the past and present, as well as the process of public engagement that is required of the Park Service when decisions are made about the property and its future. As a National Park, stakeholders for the project include anyone in the United States, not just local constituents. As one might imagine, there are many rules and regulations that factor in here. Alan did say that the Park Service strives to work closely with local stakeholders, including tribal governments and other indigenous groups.

Because the property has meaning for so many different people, it is no simple matter to conduct public engagement here. One of the Collaboratory

Art Fellows asked, “What about the people who think the Park Service shouldn’t own this land at all?” Alan did not shy away from these questions, and we had a wide-ranging discussion about the ways people attach meaning to place, political systems and the ways property moves through them, as well as how inadequate popular notions of “placemaking” really are.

Our visit to Coldwater Spring concluded with a walk to the spring, where Alan showed us photos of the site before restoration efforts began. We talked about whether or not “restoration” could ever include reconstructing buildings of this type, which, for former employees of the Bureau of Mines, have a lot of significance.

One of the Collaboratory Science Fellows pointed out similar kinds of projects elsewhere that preserve traces of industrial history, even though much of it isn’t pretty. Sometimes you have to preserve “ugly” places, not necessarily because you’re proud, but so future generations don’t forget.

Afterwards, we were able to drink from the spring—not something the Park Service recommends. We learned that Alan and his colleagues will soon begin a new phase of restoration at Coldwater Spring, and many in our group expressed interest in getting involved.

To learn more about the project and the history of area, I suggest checking out the wealth of information provided by the National Park Service on their website, as well as this list of links gathered by the Minnesota Humanities Center and Dakota creators of the wonderful Bdote Memory Map (bdotememorymap.org), which provides an indigenous perspective on places like Coldwater Spring.

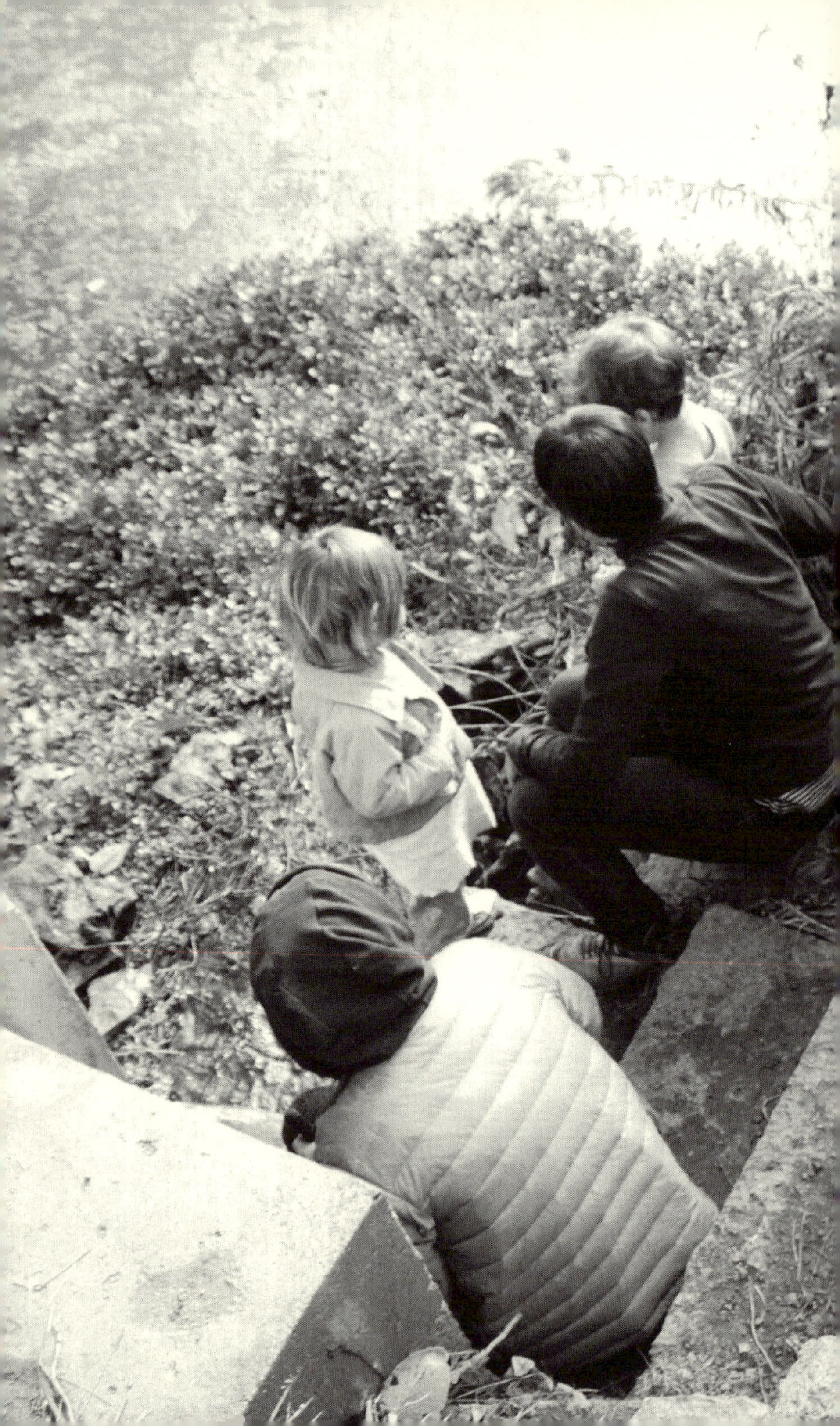

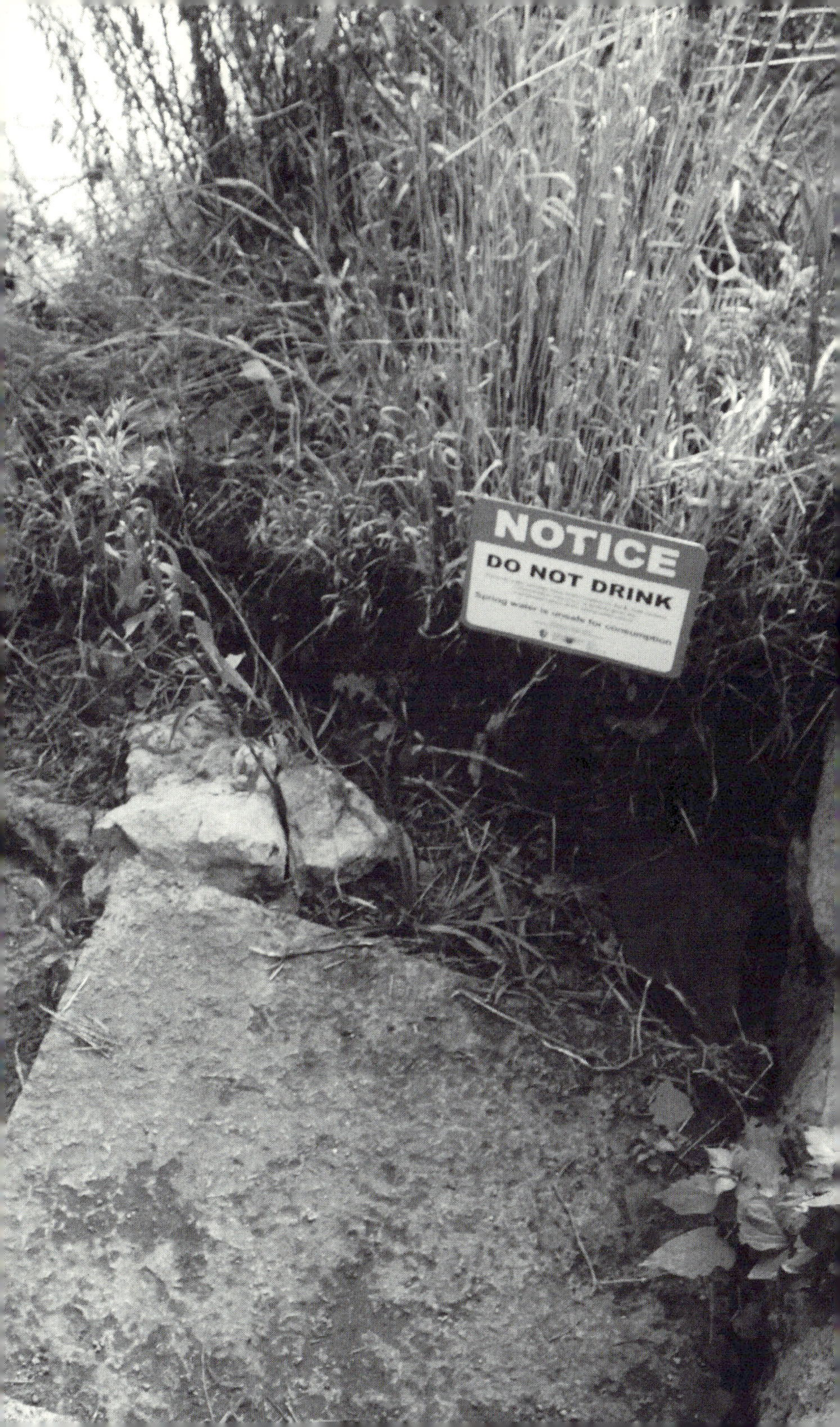
NOTICE
DO NOT DRINK

Preceding pages: Collaboratory Curator Olive Bieringa and Art Fellow Amanda Lovelee (and their children) on a field trip to Coldwater Spring in 2014. Photo: Shanai Matteson.

A Field Trip, in Truths

Dan S. Wang

Dan S. Wang is a writer, artist, organizer, and printer concerned with the relationships between art and politics, critical reflection and social action, place and history. As a member of the Compass group, a collaborative of artists and activists, Wang responds here to the City Art Collaboratory's field trip process.

My interest in science and art resonates with what philosopher Alain Badiou calls the "truth process" in his book *Ethics*. Written as the Yugoslav Wars unfolded, and particularly with the failed response of the Western powers in mind, Badiou composed his system to counter moral systems based on the so-called natural rights of man. He targets the cynical flexibility resulting from the universalizing claims of liberal humanism. With Badiou's framework in mind, I found myself intrigued by the field trip taken by City Art Collaboratory scientists and artists to Coldwater Spring, as reported on the City Art Collaboratory blog. Knowing that the area around Coldwater Spring, above the confluence of the Mississippi and Minnesota rivers, was the site of an anti-road campaign only several years after the

Yugoslav Wars, I wonder if applying Badiou's language to a consideration of that site adds productive complexity to his ethics of truth, between abstraction and materiality, in space more than time. In understanding Coldwater Spring as a source not only of water but also of truths, and considering the ethical demands it might make in turn, I hope to advance Badiou's political aims, with which I feel an affinity.

Reviewing Badiou's system is the place to start. Greatly simplified, Badiou describes the truth process as opening with the event—a specific circumstance that occurs in time, immanently rupturing a situation, and that, whatever its temporal duration, "vanishes as soon as it appears."[1] The event divides consciousness, or a historical narrative, or a demarcation between self and other, or one's relation to an imagined ideal, into the situation before and the situation after. Once perceived, the event is always already over. Those who experience the event are afterward called to observe a fidelity to it, and this is the basis for Badiou's ethics. Performing this fidelity—thereby becoming an agent of truth, touching the eternal—is the ethic of the truth process. In plain language, I take this to mean carrying forward with new being, new thinking, or changed conduct that has been demanded by the revelatory event. A basic ethical failure is to go forward as if the event had not happened. The most extreme infidelity

1 Alain Badiou, *Ethics: An Essay on the Understanding of Evil*, translated by Peter Hallward (Brooklyn: Verso, 2001), 67.

is the simulacrum of a truth, a dishonesty that legitimizes war. For Badiou, this is evil.

To support his analysis, Badiou returns repeatedly to his four-fold subjective "types," across a spectrum from intimate to public realms: politics, love, science, and art. In politics, the subject is induced through mass experience; society is the actor. Love resides at the opposite end of the intimacy spectrum, inducing a subject in the space between individual persons. Science and art serve as intermediate zones in which the individual and the social meet. To take one of the many instances where Badiou uses the fourfold matrix to illustrate his point, he lists as exemplary events in order of politics, love, science, and art, "the French Revolution of 1792, the meeting of Heloise and Abelard, Galileo's creation of physics, and Haydn's invention of the classical musical style."[2] Thus, in politics he means revolution. By love, he means the life-changing intensity of amorous relations between individuals. Science and art, activity attributable to either individual practitioners or group efforts based on shared frameworks, produce paradigm-shifting insights. But he never presents his cherry-picked examples of events as existing within layers of history, much less in relation to one another.

Moving beyond the decontextualized examples of truth events he prefers, grounding Badiou's language

2 Ibid., 41.

in a particular place reveals that different subject types may articulate truths along a common thread; in the Coldwater example that follows, in some sense all four are present. The different events entail multiple fidelities but are bound by relation of varying depths of time to the same place. Together, they produce a compound truth, at times contradictory but moving toward resolution. We could say the original truth is the spring itself, the existence of which was and is a hydrogeological event that occurs outside of human purpose. Science can make truths legible in ways that no other endeavor can, but in contrast to the theory of relativity or the invention calculus, the self-evidence of the spring precludes the need for a formally presented insight. All science of the spring, from the earliest descriptions by explorers to latter-day hydrological measurements, only supplements the fundamental fact of flowing water.

A series of Coldwater truths across subject types can be traced back to a work of art: a shimmering 1856 watercolor landscape by Alfred Sully, a commissioned officer and West Point graduate. The painting depicts the area from the perspective of the western bank of the Mississippi, above the confluence; Coldwater Spring is in the wooded foreground and Fort Snelling is off in the distance. The picture glows with summer light and the expansive skies of the northern prairie, capturing a moment when the settling of Minnesota was beginning to accelerate, and the sacred place Bdote ("where two waters come together") was subsumed by its strategic significance. As the subject type art, the watercolor preserves a truth in the image and also exists as a truth-event in its objecthood. Sully's image presents an event not as artistic innovation (Badiou's preferred art reference), but as a document of opposing vectors—that of the free Dakota lands as represented by the unfenced terrain, or the simulacrum of democracy advanced by the settler nation, monumentalized by the fort. Competing fidelities ensued, resulting in the Dakota Uprising of 1862.

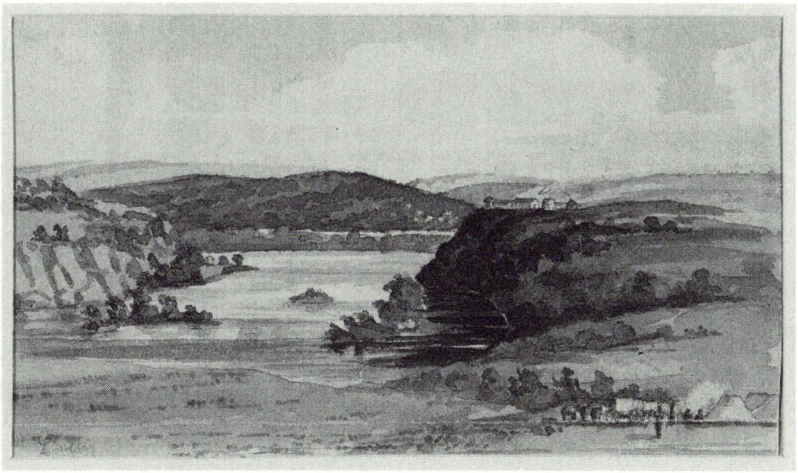

Sully, Alfred. Junction of the Minnesota With the Mississippi At Fort Snelling, View From the Hotel. 1856.

Sully seems to have observed a fidelity to the events of his own artistic creations. Even as he carried out his orders to secure territory, Sully is reputed to have been an advocate for Indians, treating them fairly and honestly, if not softly. To add to the truth in art already present in his Mississippi riverscape, an event in love seems to have emerged between the Anglo Sully and the great romance of his life, Manuela, his first wife, a Californian Chicana who died tragically young. Later, perhaps readable as a sort of fidelity to his mixed first marriage, Sully wed a French-Yankton woman in the Minnesota Territory, thereby contributing to the polyglot universe he found in the territorial settlements. One of Sully's granddaughters was Ella Deloria, an ethnographer who conducted formal research on Navajo, Lumbee, and Lakota cultures. The social scientific reports she helped author contributed to both government recognition and tribal self-recognition. These published reports were events in Badiou's sense; the reclamation and growth of tribal identities are the fidelity.

To add to these linked events in art, love, and science, and to complete the poetic redemption of the compound truth, one of Sully's further descendants, the activist writer Vine Deloria Jr., carried forth a four-generations-late fidelity to the 1862 Dakota Uprising that Sully himself played a part in suppressing. Deloria's 1969 book *Custer Died for Your Sins* helped to shape contemporary indigenous political consciousness and became a key text in the resurgence of native resistance against the imperialism of which his great-grandfather was a historical agent. Moreover, *Custer Died for Your Sins* was wrapped up in an event of its own moment—in the subject type politics at global scale, i.e., the worldwide, systemic convulsions of the late 1960s and early 70s. A generation after that, a fidelity to the heyday of Red Power was an element in the production of a new political event, one that returns the thread to Coldwater Spring.

In 1997, in response to a planned highway widening and new interchange that threatened the spring and nearby sacred stand of oaks, a resistance movement rose up and established itself physically as an outlaw road-block encampment that became known as the Minnehaha Free State. Eventually, this became the nation's longest ever blockage of a government road construction project. The campaign was comprised of those fighting to preserve the oaks, property owners resisting eminent domain, and the Mendota Mdewakanton Dakota, whose claims to tribal identity are interwoven with their claim on the spring. The Minnehaha Free State produced a militancy that kept the encampment going for sixteen months—we could say this was partly in fidelity to the indigenous rights and identity articulated by, among others, Vine Deloria. The area being defended was the geographic heart of a truth-event, fidelity to which was evident not only in the road-block, but even more so through the years of grassroots political action, testimony in public hearings, court challenges, and other kinds of dedicated resistance. That fidelity prevailed to the degree that Coldwater Spring remains free flowing, and, through persistent striving, it was eventually granted governmental protection. In grappling with the problem of ethical consistency—key to combating the moral flexibility of the neoliberal human rights frame—Badiou posits perseverance (in performing one's fidelity to an event) as the opposite of opinion. The preservation demands made by the activists of the Free State may not have been achieved in total (the oaks were destroyed), but the perseverance following the event undoubtedly exposed the original position of the state—technocratically insisting on the road and dismissing concerns about the spring, but then later scaling back those plans—as mere opinion.

The truths of Coldwater Spring described in terms of Badiou's system of ethics begins with the spring itself. Additionally, the spring exists as, and in relation to, multiple events, with fidelity to one event sometimes contributing to the production of another. It meanders through all four of Badiou's subject types, by way of bloodline or political narrative. In a constructive departure from Badiou's theory, applying his language to Coldwater Spring shows that a particular point of departure in space breaks linear time, stacking events into a single, compound truth. Furthermore, the original event, unlike the disappearing acts of Badiou's exemplary events, remains open, perpetually reinvigorated by its flowing water. Seen this way, the City Art Collaboratory visitors who quaffed clear water from the spring may yet express fidelities to the land in ways unforeseen, or may find themselves ensnared in their own Coldwater-threaded events—certainly in the science or arts of their training, but equally as likely, in politics or love. The hypocrisies of capital's moralistic cloak will never be punctured without the blades of compound truths—Badiou supplies the language, but Coldwater Spring is the pure fact and the gushing event.

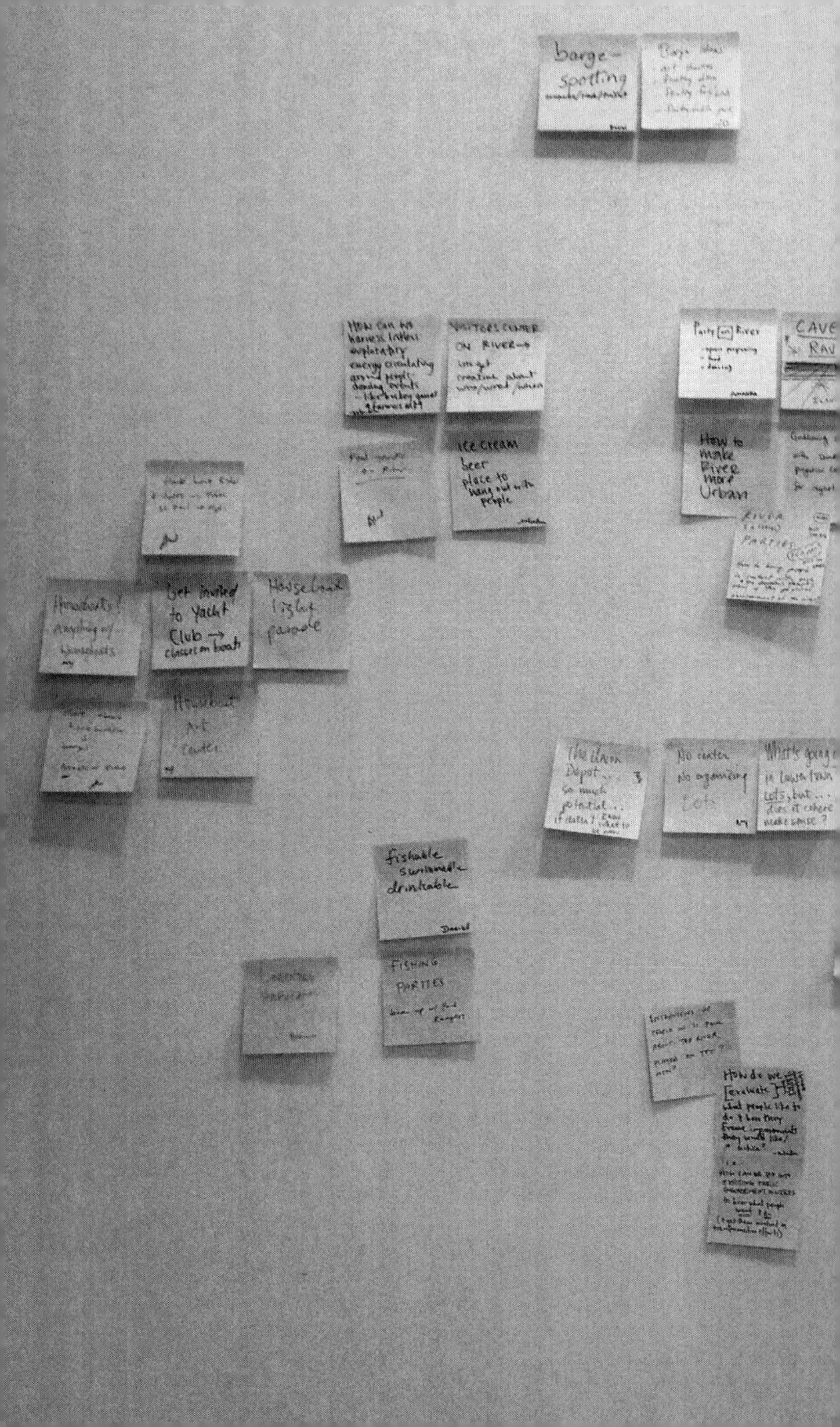

barge-spotting
VISITORS CENTER ON RIVER →
ice cream beer place to hang out with people
Party on River
How to make River more Urban
Houseboats! Anything w/ houseboats
Get invited to Yacht Club → classes on boats
Houseboat lighted parade
Houseboat Art Center
No center No organizing Lots
fishable swimmable drinkable
Daniel
FISHING PARTIES

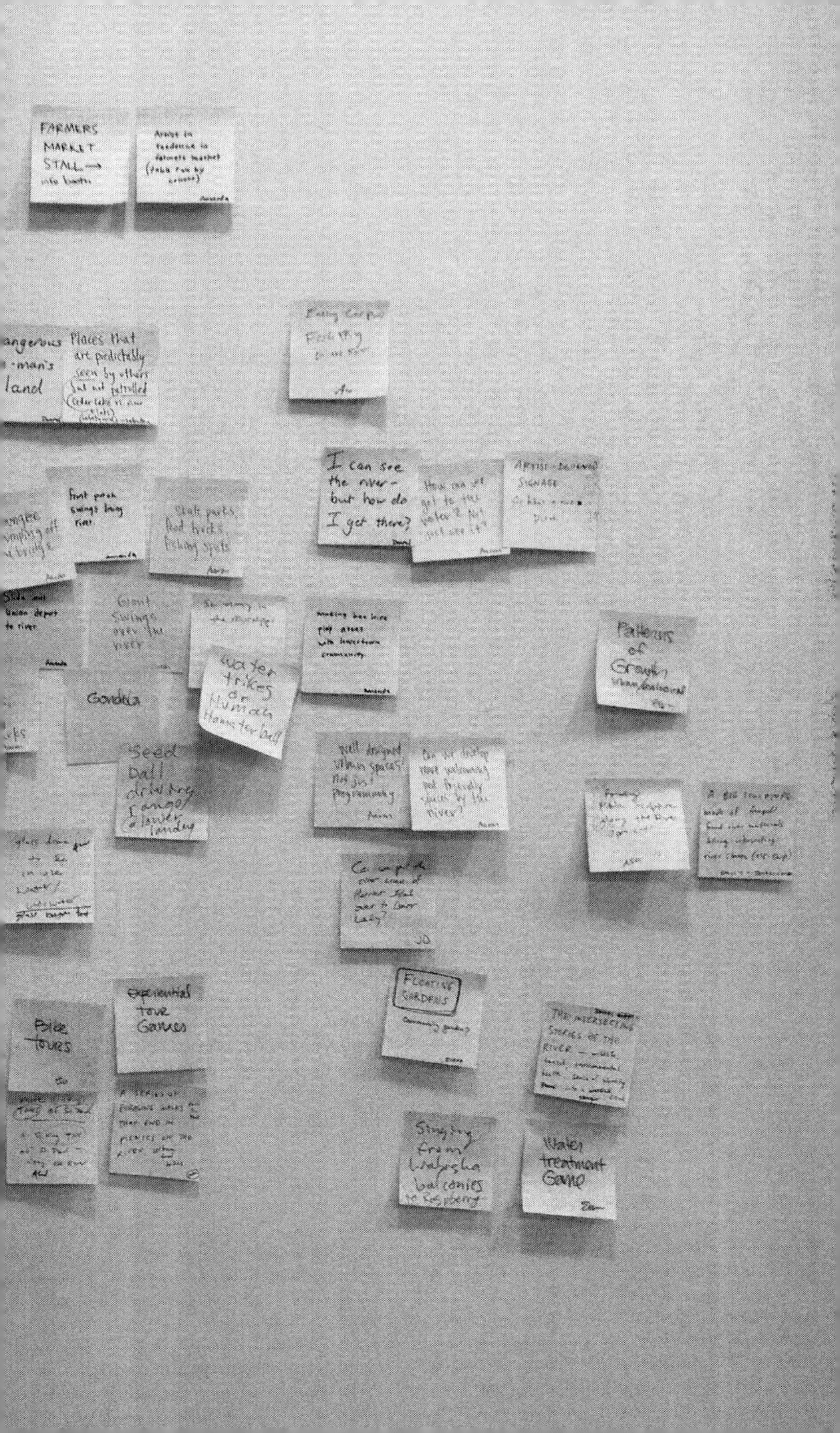
FARMERS MARKET STALL → into boats
Places that are predictably seen by others but not patrolled
I can see the river - but how do I get there?
Giant swings over the river
Gondola
Water trikes or Human Hamster ball
Patterns of Growth
Bike Tours
Experiential tour Games
FLOATING GARDENS
Water treatment Game

Preceding pages: From a Collaboratory workshop with City of Saint Paul planners and other officials following a tour of the Lowertown neighborhood and its connections to the Mississippi River. Photo: Shanai Matteson.

Relationships, Research, and Rippling Impacts

How do you evaluate the success of an open-ended program like City Art Collaboratory, which has multifaceted goals for individual participants, their respective fields of practice, and the systems on which our urban lives depend?

Though we were purposeful in our design of Collaboratory activities, we did not have predetermined outcomes such as public art and engagement projects. We did intend to build and deepen relationships: catalyzing new friendships and collaborations across disciplines, encouraging deeper and more critical relationships between participants and the places they live, and encouraging a change in perception and process within organizations where members of the City Art Collaboratory work or create art.

In early 2014, after our first year of Collaboratory activities, we decided to facilitate an impact mapping workshop for the Collaboratory cohort. This workshop, led by Eric Asboe, was loosely based on a method of evaluation called "Ripple Effect Mapping," which was developed in 2007-09 by University of Minnesota Extension educators.[1]

Asboe led Collaboratory Art and Science Fellows through a series of exercises, including one-to-one interviews, group discussion, mind mapping, and follow-up conversations via email.

From these exercises we were able to see and begin to articulate patterns of impact on individuals in the Collaboratory cohort, as well as the community of practice that was forming among the group and extending out into broader art and science communities. The following list summarizes impacts that Fellows identified with City Art Collaboratory.

Future iterations of the Collaboratory program will incorporate this technique before, during, and after the Collaboratory Fellows have spent time together.

Impact: Increasing Individual Knowledge

- First-hand knowledge about the physical environment of Saint Paul;

---

1 Scott Chazdon, Lynette Flage, Lorie Higgins, Debra Hansen Kollock, and Nathan Paine, "Ripple Effect Mapping: A 'Radiant' Way to Capture Program Impacts" in *The Journal of Extension* (Extension Journal, Inc., vol. 50, number 5, 2012).

- A fluency among artists in scientific issues and systems thinking;
- A fluency among scientists in the artistic process and systems of support for artmaking;
- Knowledge about city systems and possibilities for connection and collaboration;
- Knowledge about limits to those existing systems and programs, things that are "not working" for artists now, or that are a challenge to scientific research or education.

Impact: Increasing Individual Confidence

- A sense of agency in forming relationships across disciplines, and with the City of Saint Paul and other organizations;
- A renewed confidence in ability to effect change through collaboration at many scales;
- The feeling of being "braver together" that developed among the group, a capacity to take greater creative and collaborative risks;
- Feeling empowered (as artists) to take on roles that seem "beyond art," such as working with city planners or environmental science researchers;

- Feeling empowered (as science professionals) to test new creative approaches.

Impact: Innovative Platforms for Education & Engagement

- The group felt they'd successfully co-created a space where process mattered more than projects; where individuals could be themselves and bring existing knowledge, as well as questions;

- This "open space" proved essential for experimenting, challenging assumptions and perceptions, and finding common ground between formal disciplines;

- A space for meandering conversations, research, and intimate relationships, loosely held by an organization (Public Art Saint Paul), but not driven by a particular organizational agenda.

Impact: Network Expansion & Group Identification

- Shared experience and exchange led to friendships and broadened networks;

- The focus on process and the compare/contrast between methodologies led to stronger and more diverse networks and relationships based on an understanding of who we are as people, our passions, and our interpersonal connections;

- These new relationships also led to an expansion of resources available for the artists in the group to create collaborative work, as well as new creative resources for the science professionals in the group to access for public engagement, teaching, or developing better research questions.

Impact: Projects & Collaborations

- The group mapped a still-expanding web of new collaborations and projects stemming from the conversations and connections made in City Art Collaboratory;

- New collaborations included individual art projects, organizational education and engagement efforts, scientific research, and urban planning and design projects. Some of these projects are profiled beginning on page 141.

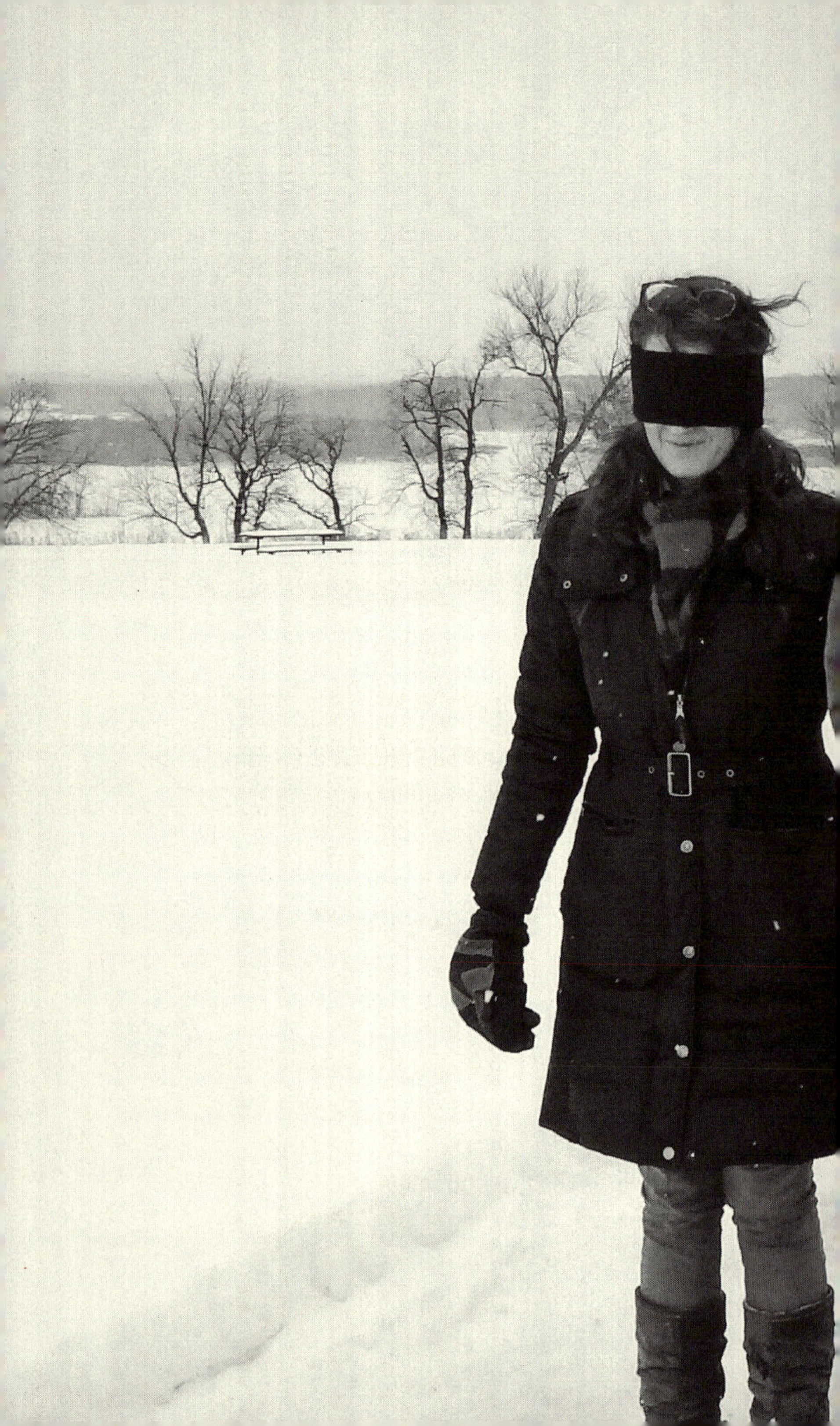

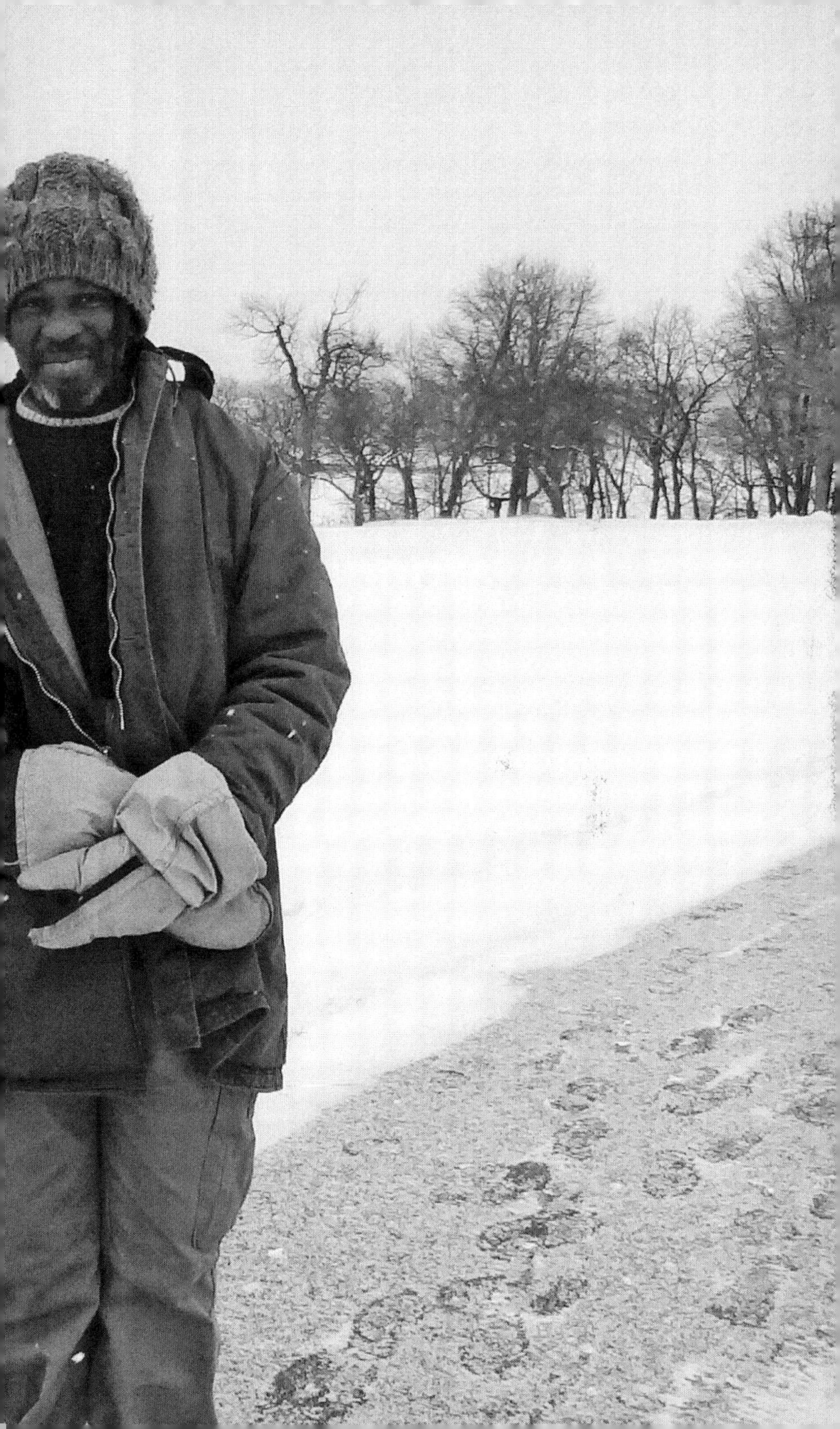

Preceding pages: Collaboratory Art Fellow Emily Stover and Science Fellow Bunmi Odumuye on a Collaboratory retreat to Schaar's Bluff Gathering Center. Photo: Andrea Steudel.

Project Profiles

In the first four years of City Art Collaboratory we decided not to prioritize the production of new artistic or collaborative public projects. We'd heard from many of the artists in our community—which has seen a proliferation in opportunities for artists to work in public or community contexts, as well as one-off placemaking projects—that what they needed more than another grant or project was the chance to slow down.

The Collaboratory is a focused opportunity to engage critically with place and practice, for artists and science professionals to deepen their work within their primary medium or community, and to develop their capacity to work collaboratively across disciplines. In particular, this work extends to new contexts in which art is now being welcomed, including within the realm of environmental science, urban planning, and community development.

Though we did not support the production of projects, we did provide a modest stipend to participating artists and scientists to be in fellowship with one another, and built other less-tangible types of support into this experience. This included the cultivation of a "community of practice" that can now recognize itself and share opportunities and resources, as well as informally advise and mentor others working in interdisciplinary terrains.

After four years of the Collaboratory, we've seen new relationships between and among the Art and Science Fellows blossom. Not only have these individuals found new opportunities outside our program, they've pushed themselves artistically and critically, at the intersection of art, science, environment, and place.

These are just some of the projects Collaboratory Fellows have led or supported in recent years, which have been selected to illustrate the variety of ways these artists and others engage with the themes and connections sparked through the Collaboratory.

The Carptactacle

Various Locations, 2013 - 2014

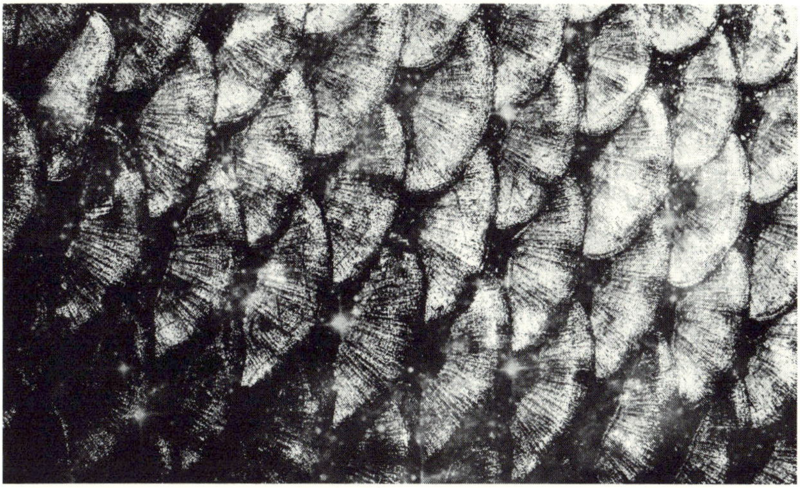

Artists / Collaborators: Molly Balcom Raleigh (Artist, Collaboratory Fellow), Janaki Ranpura (Artist, Collaboratory Fellow), Shanai Matteson (Artist, Collaboratory Director).

The Carptactacle was a spectacle of conversation, a slow-motion scientific-social happening that consisted of artistic research in search of (and resisting) a single thesis. This artistic investigation into non-native carp in the Mississippi River sifted through questions of resource stewardship, global migration, conceptions of nature, the intersections and overlaps of art and science, and the social constructions of language. Initially framed as building toward a public art convening in the form of a community fish fry, project activities included visits to points along the Mississippi, reflective conversation and writing on the development of art-science collaborations, and experimental works with invasive carp. Throughout the two-year span of the project, Balcom Raleigh, Ranpura, and Matteson presented their research as a carp dinner, live presentations teasing out public appetite for this conversation, and a short film, *10,000 Ways to Eat a Carp*, which aired on Twin Cities Public Television as part of Works Progress Studio's TV Takeover program in 2014.

Website: mollybalcomraleigh.com/carptactacle

Image credit: Molly Balcom Raleigh

Urban Flower Field

Pedro Park Site, Saint Paul, MN, 2014-2016

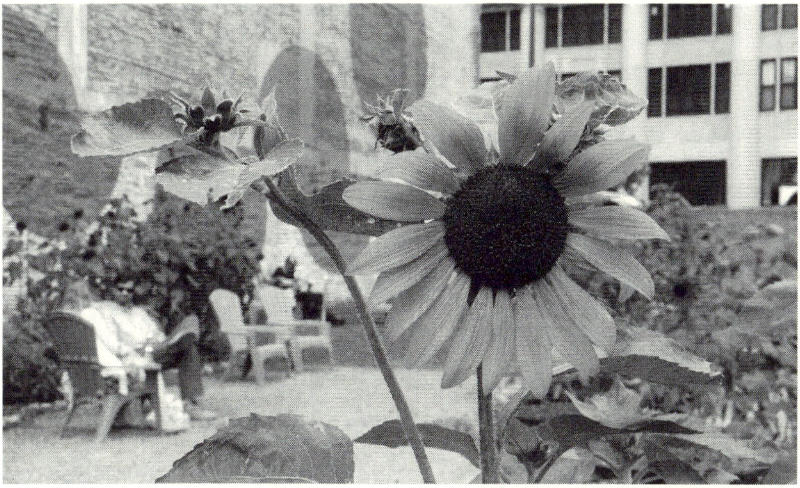

Artists / Collaborators: Amanda Lovelee (City Artist, Collaboratory Fellow), Adam Kay (Biologist, University of Saint Thomas), Don Ganje and Bianca Paz (Saint Paul Parks and Recreation), Ed Charbonneau and Jeremy Szopinski (Muralists), Paula Westmoreland (Ecological Gardens), Eric Berg.

Urban Flower Field is a temporary public park and social space at the intersection of art, science, and civic process. The collaborative team, led by artist Amanda Lovelee and environmental scientist Adam Kay, has pioneered a unique model that transforms vacant spaces, integrating scientific research and temporary public art into the process of developing community gathering space. With 5,070 acres of vacant public and private land, Saint Paul is an ideal laboratory for exploring this new process of activating and remediating urban spaces in transition. The Fibonacci Sequence—a numerical formula recognized by mathematicians and philosophers since ancient times as an order found in nature and adopted by artists and architects—serves as the design model. A spiraling set of 96 circular flower plots holds a vast experiment that tests how wildflowers take up heavy metals from urban soils. The spiral is echoed on a 4-story mural, providing a backdrop to a central patio for gathering. The space is artfully programmed with films, environmental conversations, and community artmaking, activating the space while the City of Saint Paul plans for a permanent park.

Presented By: Public Art Saint Paul, Saint Paul Parks & Recreation, University of Saint Thomas, Minneapolis College of Art & Design, and neighborhood volunteers.

With Funding Support from: ArtPlace America, Saint Paul Parks and Recreation, Minnesota State Art Board, Lunds, Minnesota State Arts Board.

Website: publicartstpaul.com/urban

Photo: Amanda Lovelee

The Plume Project

District Energy, Saint Paul, MN, Winter 2015-16

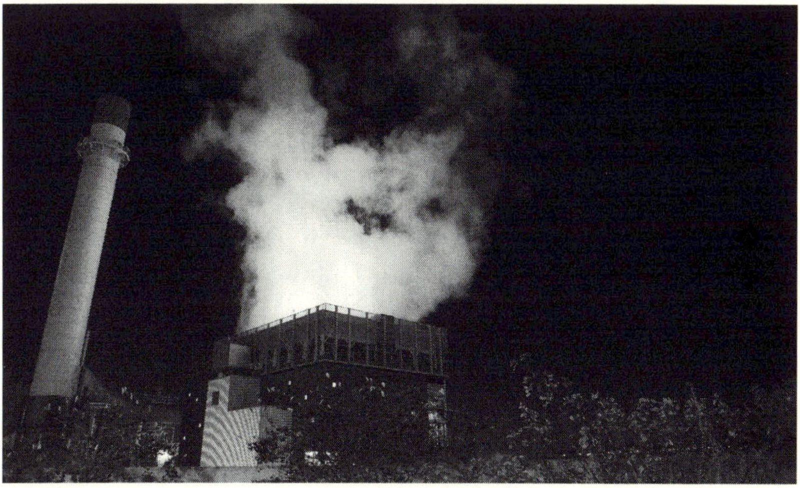

Artists / Collaborators: Aaron Dysart (Artist, Solar System, Collaboratory Fellow), Emily Stover (Artist, Rumblings, Collaboratory Fellow), Asia Ward (Artist, Coloring Contest, Collaboratory Fellow), District Energy, Andrea Steudel (Production Advisor), Kyle Waites (Lighting Designer), Max Hoaglund (Technologist), Michael Murnane (Projection Consultant).

The Plume Project was a series of temporary art works (*Solar System, Rumblings, Coloring Contest*)transforming the District Energy steam plume into a showcase for projected light displays. This utility plant, located on a main thoroughfare, works day and night in the center of the city supplying heating, cooling, and power for a demanding population. While many residents assume the emanating plume is smoke, in reality it is cloud-like water vapor from a system that primarily uses a sustainable fuel source. By shaping, coloring, and animating this plume, the artists of *The Plume Project* celebrated Saint Paul's unique efforts toward sustainable energy production, and made an often invisible urban energy system visible and vibrant.

Presented By: District Energy and the Saint Paul EcoDistrict

With Funding Support from: The Knight Foundation, Saint Paul Riverfront Corporation, Blue Cross Blue Shield Minnesota, and Ever-Green Energy

Website: plumeproject.com

Image credit: The Plume Project and District Energy

Making the Best of It: Dandelion

Northern Spark Festival, Minneapolis, MN,
June 11-12, 2016 - Ongoing

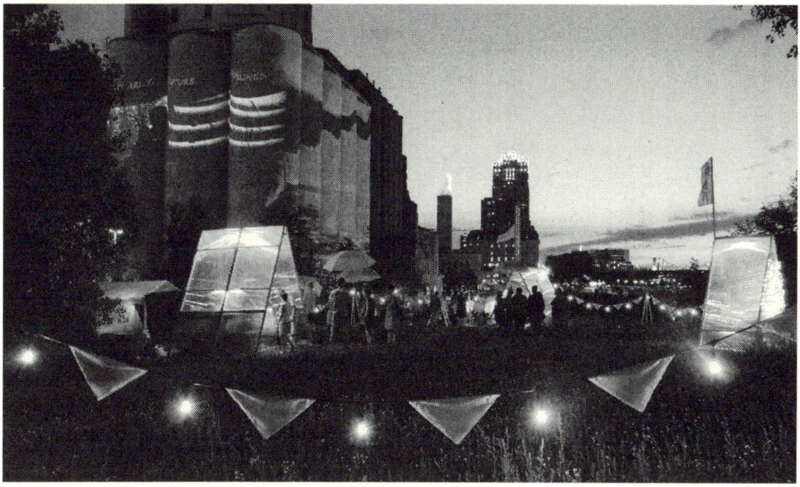

Artists / Collaborators: Marina Zurkow (Artist), Kirsten Valentine Cadieux (Artist, Collaboratory Fellow), Aaron Marx (Artist, Collaboratory Fellow), Sarah Petersen (Artist), multiple other art, science, and community partners.

Making the Best of It is an ongoing series of regionally site-specific food and art installations and community meals that feature a climate-change enabled (and often unwanted) edible species. People are engaged in tastings and conversation about the risks of climate chaos, our business-as-usual food system, and the short term food innovations at our disposal.

In Minnesota, *Making the Best of It: Dandelion* includes a team of orators, community guides, and food servers who help participants explore climate change through dandelions. A web zine provides examples of making the best of it, as our public participatory projects try to approach and question the ways in which food, behavior, and perception affect our relationships to climate change.

Presented by: Northern Spark, Climate Chaos | Climate Rising, 2016-2017

With Funding Support from: Northern Lights.mn

Website: makingthebestofit.today

Photo: Sean Smuda

In Poor Tastes Good

Art(ists) on the Verge 5, The Soap Factory, Minneapolis, MN

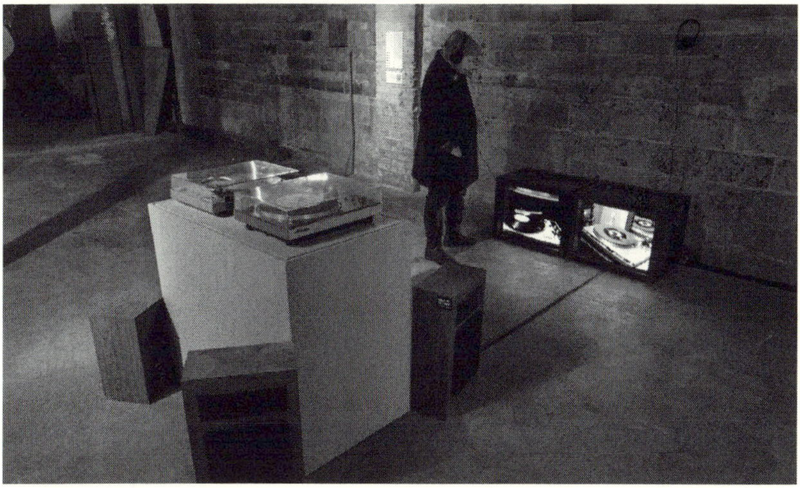

Artists / Collaborators: Katie Hargrave (Artist, Collaboratory Fellow).

In Poor Tastes Good is an exhibition project exploring the politics and history of sugar, examining moments of bitterness found therein. Artist Katie Hargrave focuses on the particular history of beet sugar, which was introduced to the U.S. as an abolitionist tactic and is the subject of recent labor struggles in the upper Midwest. Archival materials explore the transformation of sugar from sweetness to something else—a sticky mess within which we get stuck. Between metaphor and formalism, between social practice and interactivity, this project pushed the boundaries of her practice while the subject's boundaries remain almost limitless. It was part of Hargrave's *Art(ists) on the Verge 5* fellowship, and also involved other Collaboratory Art Fellows in critical and supporting roles.

Presented by: Northern Lights.mn as part of *Art(ists) on the Verge 5*

Website: http://northern.lights.mn/projects/katie-hargrave-in-poor-tastes-good/

Image credit: Katie Hargrave

Ruination: City of Dust

Lake Nokomis, Minnehaha Creek, and Minnehaha Falls, October 18-19, 2014

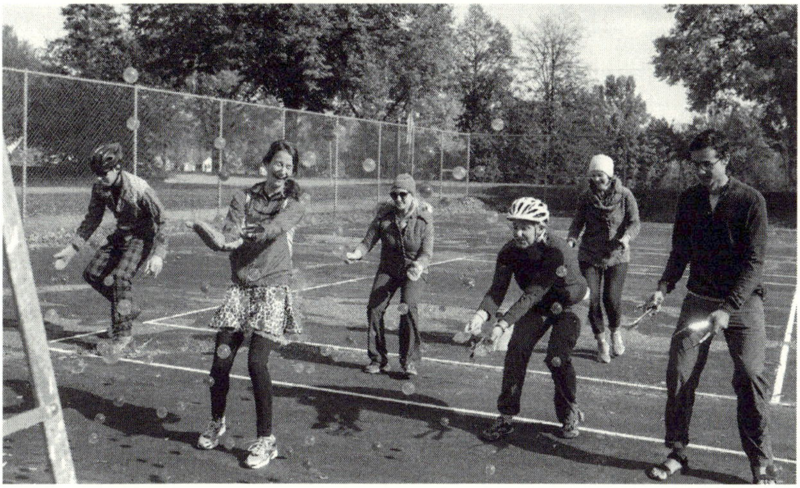

Artists / Collaborators: Northern Lights.mn, Minnehaha Creek Watershed District, Ken Eklund (Game Designer), numerous additional artists and performers, including Leslie Yetka (Minnehaha Creek Watershed District, Collaboratory Fellow) and Molly Balcom Raleigh (Artist, Collaboratory Fellow).

On October 18 and 19, 2014, nearly 400 people time traveled to the year 2314 to explore the ruins of Minneapolis, a city abandoned and covered in dust. Teams of explorers braved the harsh conditions to dig the ruins, searching for answers about which of six Deadly Problems brought the city to its final ruination. *Ruination: City of Dust* is a multiplayer mystery game played on bicycles, using the landscape of Lake Nokomis and Minnehaha Creek as a game board. Contemporary environmental science informs the sci-fi narrative, while elements of art intertwine with physical and mental challenges. This interactive, landscape-based game was developed by a creative team that included both Art and Science Fellows from the City Art Collaboratory.

Presented by: Northern Lights.mn, Minnehaha Creek Watershed District, Writerguy Ken Eklund, and Minneapolis Park and Recreation Board.

Website: ruination.mn

With Funding Support From: The Bush Foundation, through a Community Innovation Grant

Photo: Dusty Hoskovec

Water Bar

Various Locations, 2014 - Ongoing

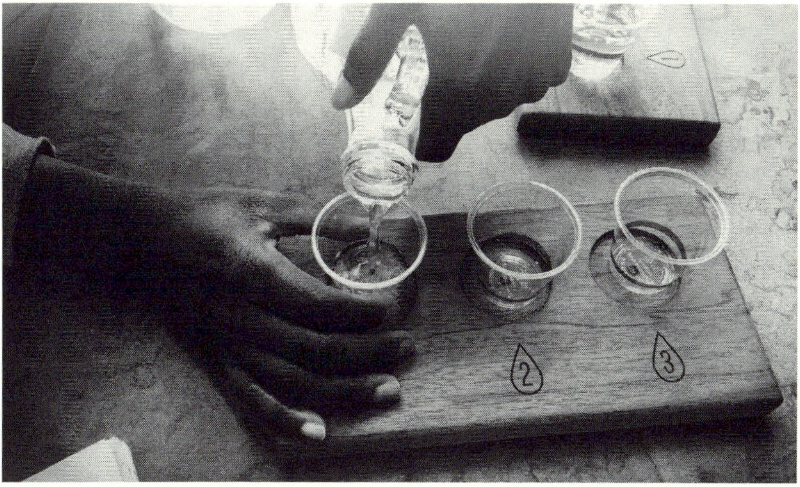

Artists / Collaborators: Colin Kloecker (Artist, Works Progress Studio), Shanai Matteson (Artist, Works Progress Studio, Collaboratory Director), Aaron Dysart (Fabricator, Collaboratory Fellow), and various other art, science, and community partners, including Kate Brauman (Scientist, Institute on the Environment, Collaboratory Fellow) and Roopali Phadke (Political Scientist, Macalester College, Collaboratory Fellow).

Water Bar began in 2014 as a public art pop-up, a bar serving local tap waters, staffed by water researchers, sustainability educators, activists, artists, and other community residents. *Water Bar* has since served local tap waters to over 30,000 people in places across the U.S., sparking conversations about the importance of local water to lives and communities, and helping to equip people to engage with complex and urgent issues stemming from our collective attitudes and approaches to environment. Issues like water pollution and scarcity, land use and community development, climate change, and environmental injustice are always on tap. *Water Bar* has also directly engaged over 60 volunteers and interns as *Water Bar* "water tenders." These everyday experts help care for and communicate about water and the ways it shapes and is shaped by our culture.

Presented by: Various partnering organizations, including Twin Cities Public Television, Crystal Bridges Museum, Illinois River Watershed Partnership, Elsewhere Museum, City of Greensboro, Cape Fear River Basin Project.

Website: water-bar.org

With Funding Support From: The Bush Foundation through a Bush Leadership Fellowship awarded to artist Shanai Matteson, various partnering organizations.

Photo: Colin Kloecker

closer

Various Locations, 2016

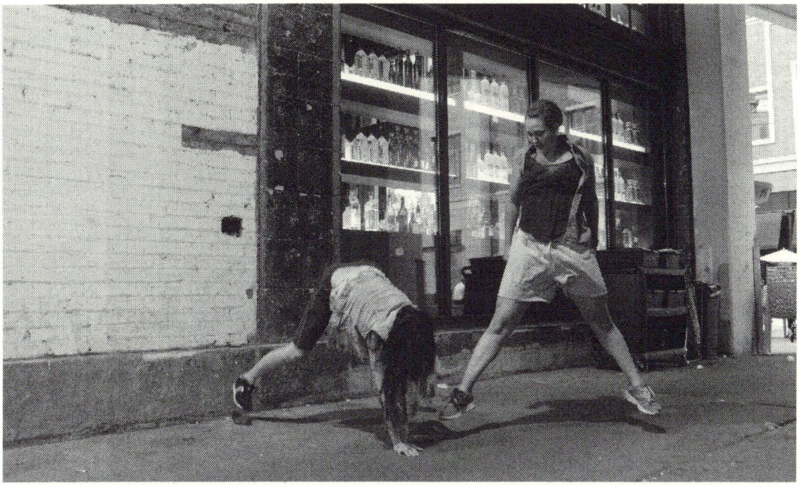

Artists / Collaborators: Olive Bieringa (Artist, BodyCartography Project, Collaboratory Curator) and Otto Ramstad (Artist, BodyCartography Project).

An interactive and intimate dance project, *closer* lays bare the power of physicality and presence through a series of one-on-one dance performances. These performances are created for one dancer and one audience, in public spaces and multiplies, becoming an expansive performance experience and dance party. Artists in the BodyCartography Project play with how the meeting between performer and audience generates the possibility for something new. *closer* uses movement to create transformation.

How do we create meaning and value around our experiences, without judgment? *closer(solo)* has been performed at Convention Center Plaza for Northern Spark, inside Minneapolis City Hall, the Stone Arch Bridge, downtown Saint Paul, White Bear Lake in Minnesota and the Wellington waterfront in New Zealand as part of the *Performance Arcade* and will soon appear in public spaces in Chicago.

Website: bodycartography.org/portfolio/closer

Photo: Sean Smuda

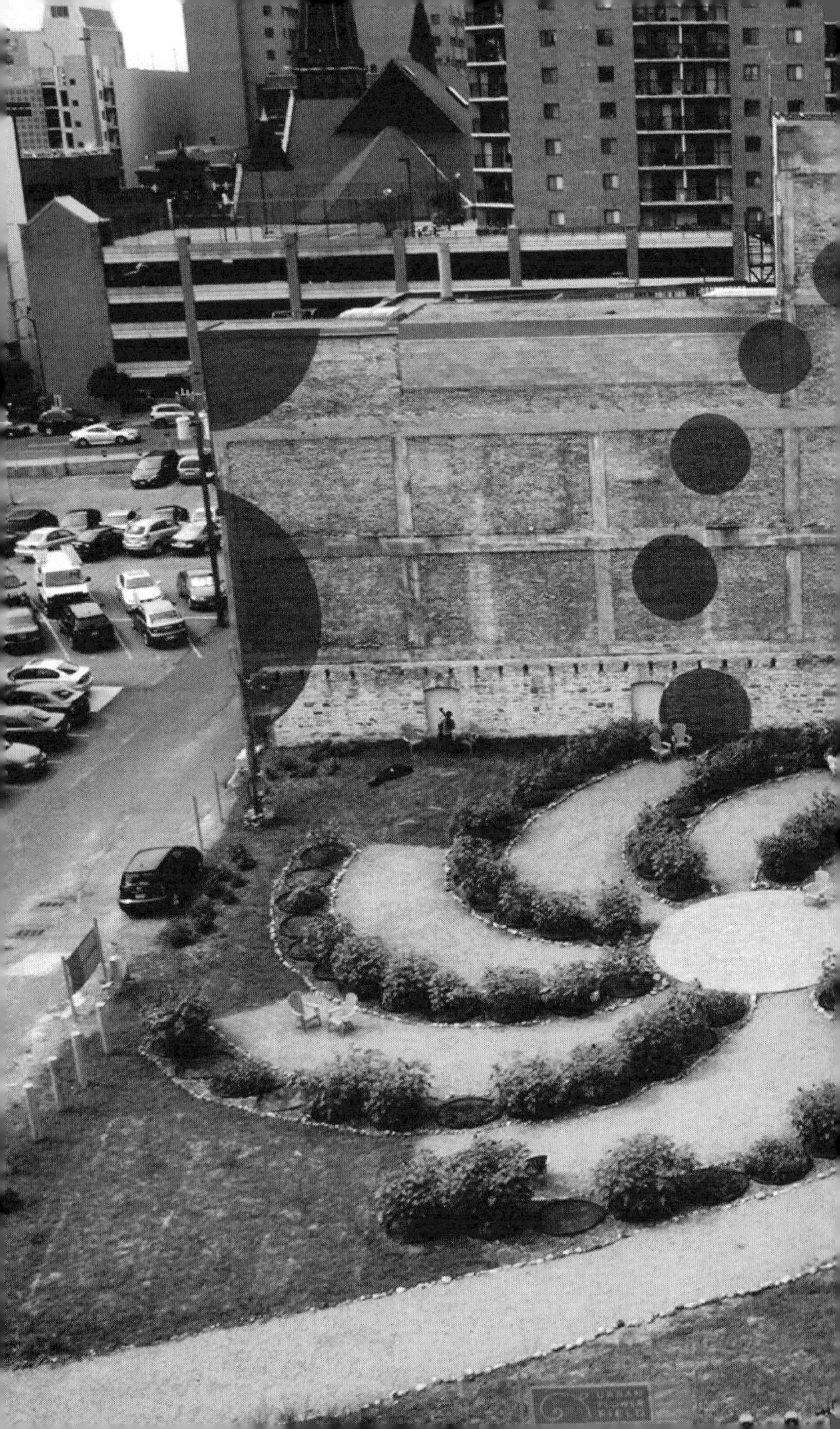

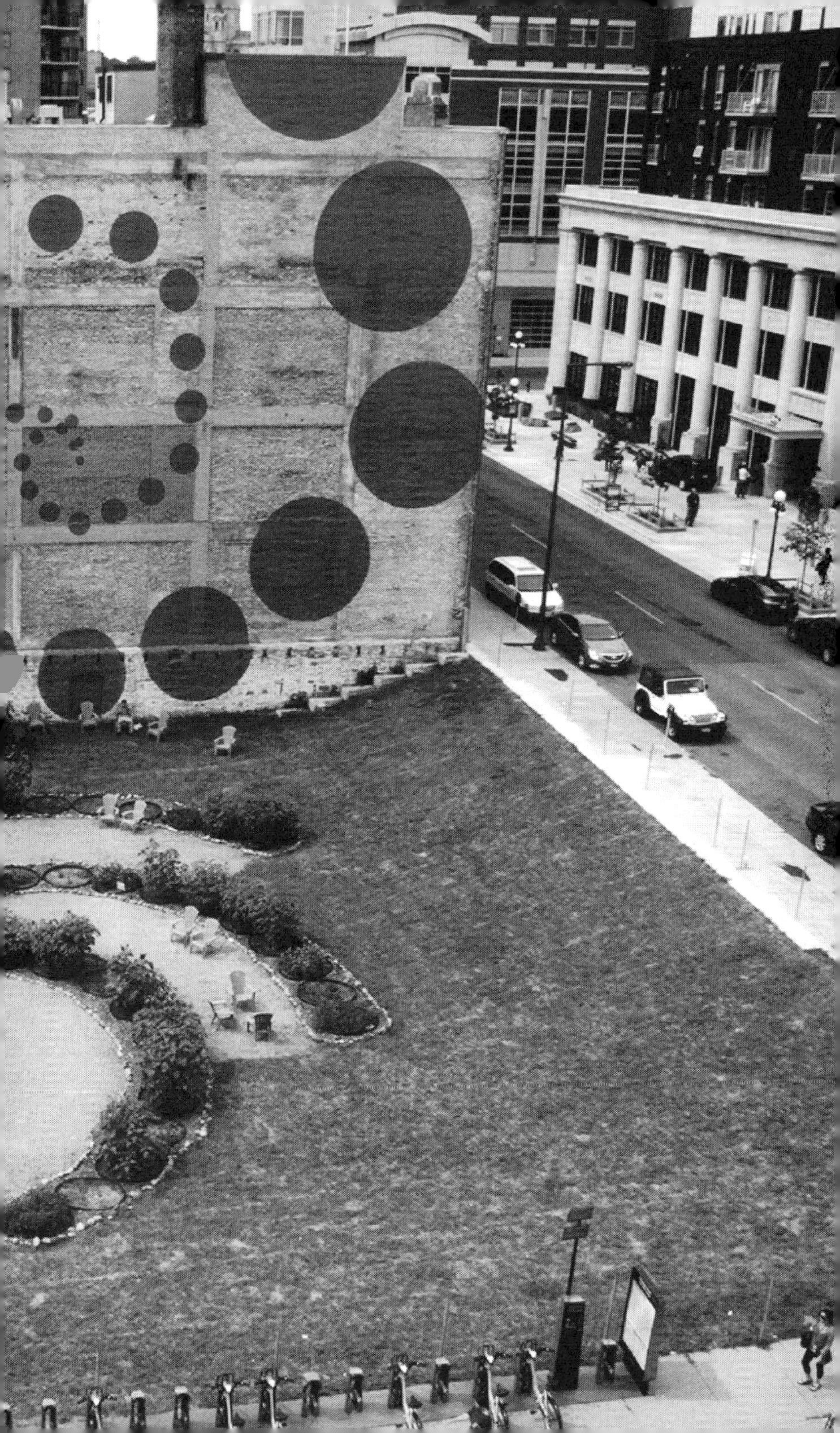

Preceding pages: Urban Flower Field. Photo: Amanda Lovelee.

Interview: Amanda Lovelee & Adam Kay of Urban Flower Field

Amanda Lovelee & Adam Kay
with Shanai Matteson

Artist and Collaboratory Fellow Amanda Lovelee became a City Artist for Saint Paul in 2013, a position of Public Art Saint Paul. In that role, she was asked to intervene in a city-owned downtown space in transition, resulting in the art-science project discussed here.

Although some cities have persisted for thousands of years, urban landscapes have always been dynamic. Patterns of abandonment and redevelopment of city spaces reflect shifting economic and social factors. One consequence of these shifts is the prevalence of vacant space within urban areas. In most U.S. cities, about 15% of land is characterized as vacant, defined by long-term underuse or abandonment. Vacant spaces are often equated with blight, but they also provide opportunities for dynamic experimentation, to make urban areas vital and enriching for all residents.

Working with a vacant or underused space, Lovelee and Kay wanted their collaboration to contribute new information about how artists and scientists can work together to transform an urban space into a public good. *Urban Flower Field* became an exemplary project for this vision.

They created *Urban Flower Field* on ¼ of a city block in a commercial and growing residential section of downtown Saint Paul. The land was the site of the Pedro Luggage Company, until the owners donated the property to the City of Saint Paul in 2011, with the stipulation that it be converted into a park. The site remained largely undeveloped for several years after the building had been demolished, and nearby residents started to get upset that the promised park was still a fenced-off gravel pit. The City's Department of Parks and Recreation asked City Artist Lovelee to create a public art installation that temporarily transformed the site from an abandoned lot into an attractive community gathering space. Lovelee approached Kay, a biologist at the University of Saint Thomas, to create a unique hybrid art-science project designed around an urban ecology experiment.

Their experiment resulted in *Urban Flower Field*, embedding into public art an overall design based on the Fibonacci sequence, a mathematical spiral that is found in both nature and art, thus linking science and beauty. The science experiment explores how plants can reduce contaminants in urban soils. The "test subjects" are beautiful wild flowers, including Black-eyed Susans and Crimson Thriller sunflowers. The garden plots are arranged in eight spirals that extend out from a circular central patio. Fieldstones, painted by community members, line each of the spirals. Artistic signs and paintings explain the artistic vision and the science experiment. Colorful Adirondack chairs provide places for visitors to sit. The landscape design is echoed in a vast mural painted on a brick wall bordering the site. The mural was created by Ed Charbonneau and Jeremy Szopinski. A project of this scale could only be realized by a large team, which also included Don Ganje and Bianca Paz from Saint Paul Department of Parks and Recreation; Hunter Gaitan and Liz Scherber as University of Saint Thomas interns; Abby Kapler as a Minneapolis College of Art and Design intern; Paula Westmoreland from Ecological Gardens, Eric Berg from Berg Brothers, many neighborhood volunteers, and the support and collaboration of staff at Public Art Saint Paul. Support for the project came from ArtPlace America, the University of Saint Thomas; Saint Paul Parks and Recreation, the John L. and James S. Knight Foundation; the McKnight Foundation, and Lunds & Byerlys grocery stores.

The place created through this melding of art and science has encouraged community interactions. The site is a public space, where residents, working people, and visitors often gather informally. Public events at the site, including film showings and art-science discussions, have further facilitated community engagement.

An informal survey revealed that visitors think the project has transformed the space from "unhappy" to a "lovely place that serves as a green community gathering space." The aesthetic and conceptual aspects of the site led to a 2014 "Great Places Award" from the Sensible Land Use Coalition, a regional organization that promotes urban development.

Shanai Matteson: *How does the project fit into your respective practices?*

Amanda Lovelee: As an artist, I am interested in the places where people gather in contemporary society and how they connect. I build tools/spaces to bring people together, to have conversations, to make the invisible visible, to raise awareness and create change. Working as a City Artist for Public Art Saint Paul, I view my job as a translator between the city's ideas and residents' hopes, with the goal of building a city everyone wants to live in. We have a long list of urgent issues in city building in the 21st century around ideas of sustainability, equity, health, education, housing, infrastructure, and safety. This list feels overwhelming, and not one sector alone can conquer these issues.

I want to make my neighborhood, my city, and the world a better place. But I have sometimes struggled to find how I can use my skills as an artist to truly create positive change. Upon graduating from art school, I quickly realized I did not want to only make art that you need an MFA to understand. The gallery system was not something I was interested in. I never know what box to check to describe my medium, or the best way to explain my practice. I really connected with Jen Delos Reyes's words about social practice art: "It is about the value of art in daily life and the belief that art is for everyone, not just the elite. Its commercial value can be slim as much of the time this kind of work might not look like art at all. This work promotes agency in artists, it is made alongside and with its intended audience and necessitates being in the context of the world."[1]

When the job of City Artist was posted, I was ecstatic. It was my dream job. That was 4 years ago, and it has been my dream job, struggles and successes included. *Urban Flower Field* was the first project I did as a City Artist. It has raised the bar on how I think about City Art and its potential. The project pushed me and is still pushing me. I never thought I would find the sweet spot in my practice where I could combine my interest in aesthetics, urban design, gathering people, dreaming big, and the natural world. There are still days where I just walk over to *Urban Flower Field* after a long meeting to watch people enjoying the space and remember why I do what I do. *Urban Flower Field* did not necessarily fit into my practice but expanded my practice. The experience taught me the power of

1 Jen Delos Reyes quoted in http://badatsports.com/2016/we-woke-up-like-this-reflections-on-social-practice-in-the-sunday-new-york-times-2/.

successful collaboration across art and science and that artists can be leaders in the creative re-imagining of our cities.

Adam Kay: I am an ecologist interested in understanding how urban areas can become more environmentally sustainable and healthy for their residents. Over half of the world's population currently lives in urban areas, and more than 3 billion additional urban residents are predicted by 2050. Urbanized areas already produce most of our carbon emissions that contribute to global climate change. Ecological research can help us shape urban areas to reduce our environmental impacts, and it can strengthen the connection between city residents and nature. Unfortunately, there is not a lot of ecological research occurring in urban areas. Part of the reason for this is the lack of access to land for field sites.

Vacant spaces could be ideal for conducting urban ecology research, but there are drawbacks. Ecological research sites could help create urban greenspace while providing valuable information. However, the idea of setting aside land for research may not appeal to community members who would like to get more tangible benefits from land that is currently vacant. I think the *Urban Flower Field* concept provides a new way to break this trade-off between studying ecology and providing immediate community benefits. By embedding our experiment in public art, we were able to collect valuable ecological data while creating a site for local residents to enjoy and build community. I am hoping that this concept can help advance the field of urban ecology. I think it can be replicated in a variety of flavors in urban areas all over the world.

SM: *What were some of your big insights and challenges?*

AL: That artists can change the world, but they cannot do it alone. Cross-sector collaborations and creativity are necessary to rethink issues within our urban cities. That is my large, lofty insight, but I also have many everyday project insights/challenges.

I think a challenge is how to tell the story of this project. It involved so many people and years of working together. When you visit *Urban Flower Field*, it feels like a nice urban oasis with flowers and chairs among a sea of pavement. The people sitting in those chairs might not know that this project comes out of a system-focused idea called "City Art." The idea is to create art within the city systems—just like how the snow is plowed. We believe that art is just as important to the everyday life of residents as the sewer system, and we need both. When sitting in the park you may not realize that the large sunflowers shading you are also cleaning the soil you are sitting on. Or that much of urban soil is contaminated by decades of activity that has happened upon it. You might also not know that the space has won awards for its innovation, yet it was made on a shoestring budget. Or that the young man mowing the lawn is actually a biology student also testing the soil. We have not figured out how to share all the small details that make this space so unique, and how we believe that collaborative art/science/civic projects like this could really transform cities for the better.

We also have to sometimes fight the ridiculous battle of why this project is art and why this project is science. By bringing our fields together, there are some who view this as a weakening of each discipline instead of seeing the intrinsic value the project brings to both fields. If we want to tackle the large issues weighing cities down, then art and science might not look like they have in the past. As social practice artist Theaster Gates said about art in a recent talk at the Minneapolis Institute of Arts, "Be more purposeful and masterful with what we have. Yield more then you need. It's like gardening."[2] We believe that these kinds of daring, collaborative projects could grow new ideas and hope.

AK: A main insight from our project is that our concept can work. We were able to pull off a hybrid science-art project in the heart of a downtown area. We were able to simultaneously collect scientific data, beautify the site, and engage the public. We hope this success provides opportunities for us and others to apply this concept to other vacant spaces.

---

2 Theaster Gates, "The Art of Building Community," Minneapolis Institute of Art, February 11, 2016.

Another insight came from the science research, which provided some support for the hypothesis that plant biodiversity can help reduce soil contaminants. We are currently preparing this work for publication in a science journal. It will be exciting to share these findings with the scientific community and the general public.

The main benefit of our project—an opportunity to engage with the general public—also presented a significant challenge. Ecological experiments are typically isolated in part to avoid disturbance by people, but public art installations are designed to encourage community gathering. At *Urban Flower Field*, we created a central patio and provided portable chairs to encourage people to visit the site and socialize. We did not use fencing to protect our experimental plots because we wanted everything to feel welcoming. As a result, we had people and pets occasionally walk through (and pee on!) our experimental plots. We overcame this disturbance by making a heavily replicated study that allowed us to show patterns in the face of uncontrolled variation.

Another set of challenges stemmed from our need to make the site look nice. Features that may have been valuable to the experiment, such as bare-ground control plots or the use of weedy species, would have been ugly and not appealing to the public. We also had to put a lot of effort, time, and money into maintaining the site for continuous public presentation. It was fun to do this, but time and money are always limited, and we would not have had to worry about aesthetics if the site was isolated.

SM: *What was it like working together across disciplines and in public?*

AL: Working in the public realm, the fear of failure was high. The work I do as City Artist represents the City, which means many people are armed and ready to critique it. I also represent the idea of an artist. I do not want to fail in the name of all artists or as the City of Saint Paul. When I make something in my studio and it might not work, I move on. No one will ever see the project. I really wanted to show the city and residents that artists are great leaders, innovators, and hard workers, and that every city needs an artist on staff.

Working together across disciplines really made this project possible in multiple ways. It diversified our ideas and strengthened them. We built something together that neither of us could have built on our own. We pushed against what is considered art and what is considered a research project. It was not always easy because the collaboration took blind trust. I had to trust that Adam knew what he was doing and could pull it off. I also had to show this confidence in all conversations with the city and my own organization even when I could not speak his language. Adam had to trust that I could run a project of this scale, keep each part of the project moving, and that my overall vision was going to be great when he could not speak my language.

In the end, we both learned a new way of speaking together. Whenever the project had challenges—lack of funding, grass not growing—or successes, we stood together. Adam has big vision and an even larger drive to get it done. This collaboration and project gave me hope for the future!

AK: Science and art are not as different as they are typically presented to be. Science experiments have to have a carefully planned design and protocol, but so do art projects, particularly public projects done in collaboration with a city government. Scientists also have to be creative. In ecology, there is a premium placed on novel contributions, that is, your study has to be new to the field, and it has to build upon previous work. You have to be creative to come up with something that will take the field in a new direction while providing useful information to the general public.

That said, Amanda and I certainly could have had problems creating this project. Our goals were somewhat different. I was looking for a way to test a hypothesis in collaboration with my students. Amanda was tasked with creating a space that could help transition Pedro Park into a community asset and, eventually, a permanent park.

Luckily, we worked together very well. Let me give an example. When I told her about our idea of an experiment to test how plant diversity affects soil contaminant level, I proposed that we create square experimental plots in a grid—perfectly symmetrical and regularly spaced. Very practical. Amanda suggested that it might be more engaging if we used circular plots in the spiral design that you see at the *Urban Flower Field*. Obviously she was right! In general, it was easy to follow Amanda's lead on the aesthetic issues and on many other things. She is so creative, passionate, and earnest. I knew I could trust her instincts.

SM: *Where do you hope to take the project and/or collaboration?*

AL: The idea of *Urban Flower Field* was that it was just the first in many cross-discipline art/science/city projects, growing a new temporary vision for these spaces in the transition across Saint Paul. I am excited to see what space we get to transform next, what community we get to work with and what we grow. I hope that *Urban Flower Field* is just the first of many projects that I get to work on with Adam. I hope that we have more meetings in green houses, over-caffeinated conversations, and an even longer list of dream projects than we already have. I believe that *Urban Flower Field* can be used by other cities as a way to transform a community and spaces in transition temporarily while the long term planning for the space is happening. *Urban Flower Field* has shifted the neighborhood's perception of the space from an eyesore to a place of beauty, to their park. As we shift into the next phase of planning for the permanent park that will exist in this space, we now have an engaged and excited community who feel ownership and love for the space. If one single project can advance scientific research, ignite a community, create a beautiful refuge within a city, and increase civic engagement, imagine what 10 of these projects across the city could do!

AK: I am really interested in using this concept in urban agriculture. My students and I have been working on the biology of urban farming in traditional community garden settings. I think it would be amazing to have an *Urban Flower Field*-type project that is built around an urban agriculture experiment. We could have a design that includes picnic tables and food trucks. At the same time, we could be conducting research on composting, mulching, or some other understudied aspect of urban farming. One specific idea would be to focus on transforming paved areas that have been abandoned. They can be such a blight on a community. We could build raised beds in areas like that and use those beds to test hypotheses. That type of project really could meet multiple community goals.

I would love to do this and similar projects with Amanda. She has helped me see so many new possibilities, and I have really strengthened my feeling of connection to the community.

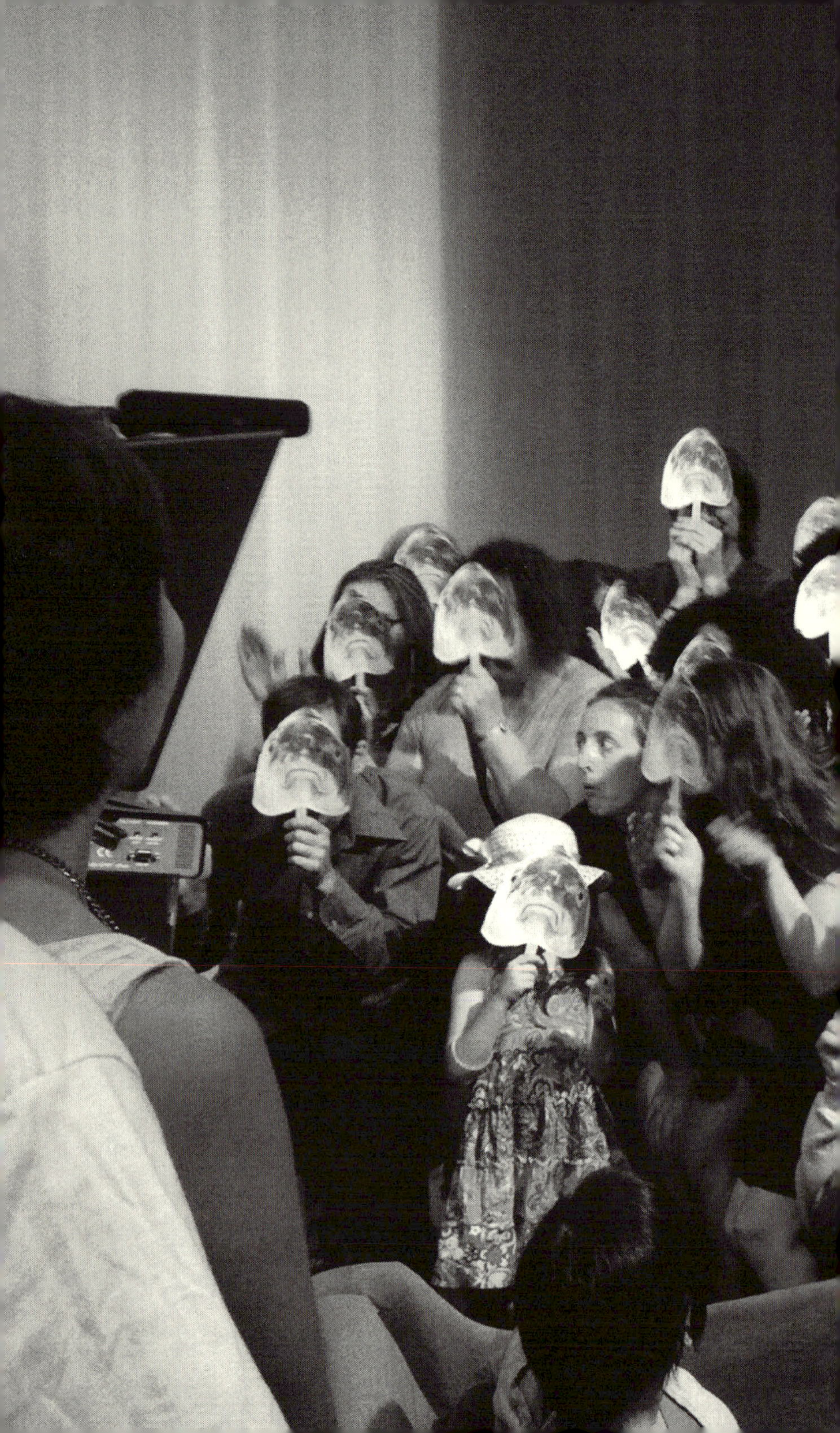

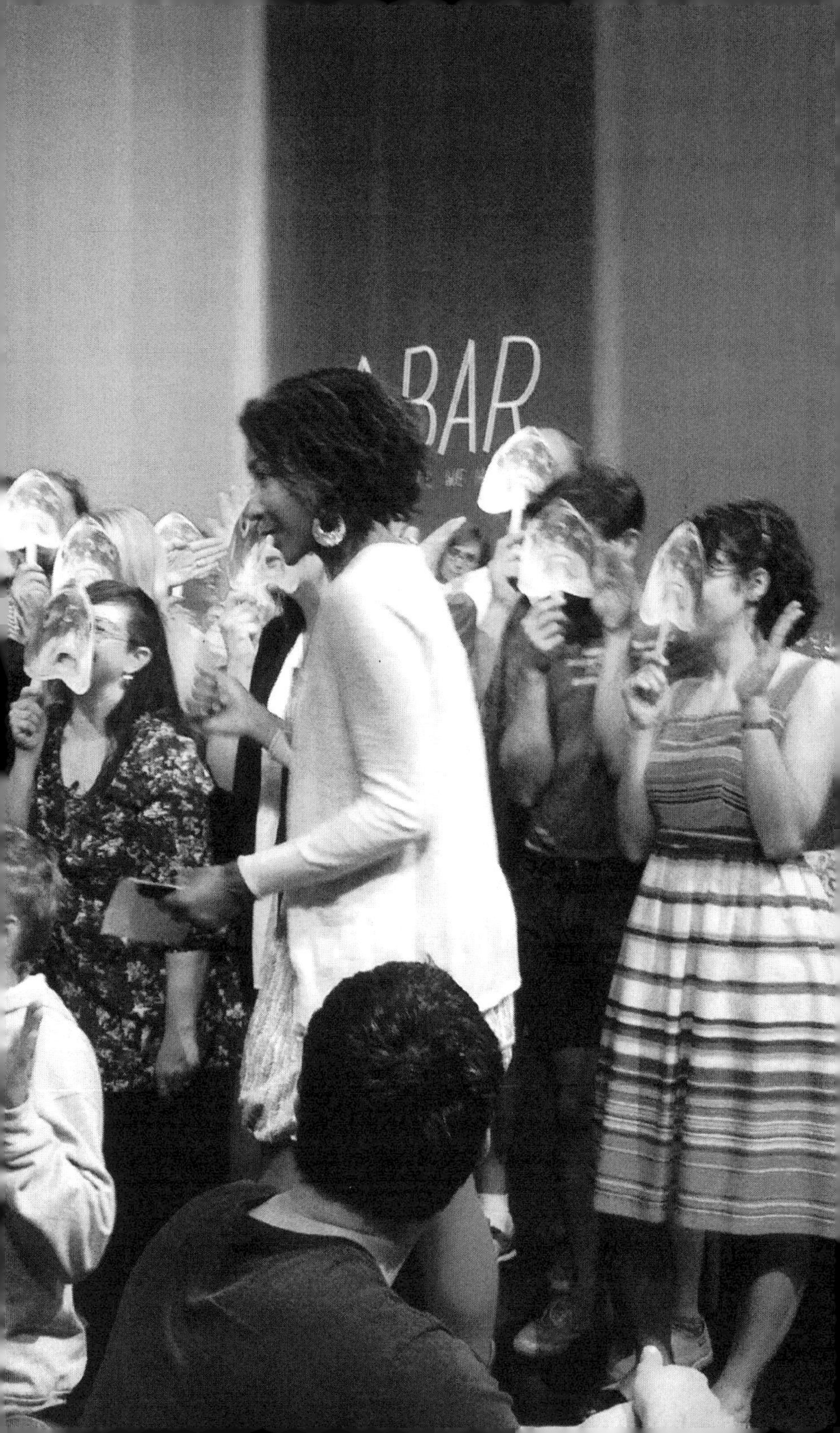
BAR

Preceding pages: Image of The Carptactacle at a TV Takeover program curated by Works Progress Studio that aired live on Twin Cities Public Television in 2014. Photo: Christine Baeumler.

What We Talk About When We Talk About Carp

Molly Balcom Raleigh

Collaboratory Art Fellow Molly Balcom Raleigh reflects on her artistic research process, following a line of inquiry about a common fish through science, culture, public engagement, and natural resource management.

The Carptactacle was a spectacle of conversation, a slow-motion, scientific-social happening that consisted of artistic research in search of and resisting a single thesis. Our investigation into non-native carp in the Mississippi River navigated questions of resource stewardship, global migration, conceptions of nature and place, and the social constructions of language, among other topics. Our activities included visits to relevant points along the Mississippi, reflective conversation, written and object-based documentation of our developing collaboration, and the handling/mishandling of the bodies of a notorious gang: the four species of Cyprinidae that comprise the so-called Asian Carp.

We initially framed our research to build toward a spectacular public art event with performances and projections—a big community fish fry. While we never arrived at that particular artistic endpoint, we found much to inspire and challenge us along the way. We shared our research through live presentations and a short video piece. This essay attempts to document our process (from the perspective of one collaborator), and to capture some of our learning as we tugged at the overlapping edges of science, public engagement, resource management, and artistic research for nearly two years.

How did I get to carp?

For me, invasive species as a river issue became an access point to the Collaboratory. Prior to this, my own interests led me to investigate conceptions of nature and meanings of place. This project was an opportunity to continue to explore ideas that are core to my intellectual and personal concerns in a way that was focused on experiential inquiry, sensory awareness, and social making.

I came to nature as a subject through food and the performative body—eating as a way of knowing self and creating relationships.

I began thinking about nature and belonging in my early 20s when I found a photograph of myself as a little girl in Michigan, sitting on the hood of my parents' station wagon, surrounded by a harvest of morel mushrooms. Picking morel mushrooms was fundamental to our family's mythology of central Michigan—where my parents are from and which we left not long after that photograph was taken. Tied to memories of my grandparents and their friends, my parents' Michigan was rural, filled with wild places where they fished, hunted, picked berries, and foraged for morels. I remember the buildup to picking wild blueberries as a 3-year-old, fantasizing about them and the intense anticipation of how delicious they would be. When we finally found and picked them, I was as intensely disappointed by how small and sour they were.

As a family, we hadn't done this kind of foraging for decades until I asked my parents to consider trying it in Minnesota. The joy of finding and eating morels again was vivid for them, and was an important early point of connection as we rebuilt new bonds as adults after years of strained relations. Foraging for morels in Minnesota was a many-layered homecoming, and at the same time, an awakening.

From morels, I widened my explorations to other mushrooms, and then to wild greens, flowers, and other plants. I shared my enthusiasm and new-found knowledge with friends. During this time, I was working on research into the intersections of food and performance, and how eating could operate as a text in performance.

Wild food foraging emerged for me as an embodied way of understanding place, and a way of sharing, sustaining, and growing this understanding with others.

Wild food captures the terroir of place. To eat of a place is to know it in a particular way, through the body and the intuitive cognition of the senses. In observing, identifying, researching, and experimenting with the edible wild greens and mushrooms of our region, I learned about some of the indigenous practices with these foods, including many plants and mushrooms that I had previously considered inedible. I also encountered the natural resource management practices of our region, particularly the classification of some plants and animals as invasive.

At the time that I applied to participate in the Collaboratory, I was following a growing gastronomic mini-movement called invasivorism, wherein chefs and conservation officers worked together to develop recipes that reframed invasives into gourmet ingredients. The idea is to increase the "natural" demand for these foods and use this demand as a means of abatement—a market solution to a market-created problem. Their motto is "If you can't beat 'em, eat 'em." To me, this was an alluring proposition, and it influenced my thinking about invasive species at the time of our first field trip to Lock and Dam No. 1.

How did we get to carp?

On a cold, bright January morning in 2013, I stopped at Kopplin's Coffee in Saint Paul to pick up Janaki Ranpura, my traveling companion for the day. We had only recently met as newly-minted Fellows of the City Art Collaboratory, on our first field trip to Lock and Dam No. 1. We had both been intrigued by the issue of invasive carp that framed our visit to the dam. With an open-ended invitation from Public Art Saint Paul to follow our own interests in the Collaboratory, we decided to try to get our heads around this carp issue by putting some in our stomachs. The invitation to prepare a few carp dishes for the next Collaboratory gathering was on the table, and I eagerly took it up. When I sent word to the group that I was driving to pick up a fresh grass carp from a fish processing plant in Pepin, Wisconsin, Janaki offered to come with me.

In Pepin we found a window onto an entire regional industry focused on catching, processing, and shipping "less-desirable" fish from the Mississippi River. Many of these fish are considered nuisance species, at least two are categorized as harmful invasives, and few if any have conservation limits. Schafer Fisheries, based in Illinois, buys huge quantities of fish from hundreds of commercial fishermen up and down the Mississippi. They process them into flash frozen cuts, smoked fish, kosher meat substitutes, and liquid fertilizer. These products are shipped around the globe to markets in Scandinavia, Israel, and Central

America, among others. This means that some fishermen export their local sea bass or snapper to be served in Minnesota restaurants, and go home to a meal of much less expensive imported carp or drum from the Mississippi River. The absurdity of this loop illustrated how the perceived value of certain species can dislocate resources in globalized cycles of poverty and power that dishonor or ignore cultural contexts.

The workers at the Pepin plant enthusiastically showed us around and we talked with them about the Collaboratory. When we mentioned we were thinking of making a video, they said no problem—they were already stars in a reality TV show made by the Outdoor Channel, "Bottom Feeders," (similar to Duck Dynasty or The Deadliest Catch). Apparently it was a sensational idea that a culture of fishing could emerge around these fish. We took photos and promised to keep in touch as our research progressed.

Back in Saint Paul at a Collaboratory event hosted by Public Art Saint Paul board members, we served our enormous grass carp whole, steamed and seasoned with ginger and scallions, along with smoked common carp and carp hot dogs created by Schaefer Fisheries. Something about serving and eating this humble, maligned fish in the plush and proper environment of a Summit Avenue mansion took hold of our imaginations. We sensed this could be an opening onto critical conversations about belonging, value, power, and collective visions of and actions on the future of our region. We also noticed how the language around the invasive carp issue, which at that time was the focus of several massive state and federal budget requests, shadowed another area of tense national conversation—illegal immigration.

At the Collaboratory pitch night in November 2013, Janaki and I presented a sketch of our project: a carp fish fry hosted in a space filled with projections, installations, and performances. The experience would be threaded together by guided conversations with field experts and other researchers. With this basic framework, we invited Collaboratory Director Shanai Matteson to join us as an artistic collaborator. Together we created a wide canvas of critical questions and ideas, bits of found poetry and metaphoric imagery from mythology, culinary history, economic theory, conservation science, and countless YouTube videos about carp. After many conversations, research, and a field trip back to Pepin, we made a short video that re-cast carp as a metaphorical protagonist in our collective inquiry. It aired on Twin Cities Public Television as part of Works Progress's "TV Takeover" in June 2014. *The Carptactacle* came to a close at the end of 2014 as Janaki and I each moved abroad.

My account of this work makes it seem logical and orderly, but in truth these realizations and decisions came about through a flood of other conversations, ideas, set-backs, and consuming projects. The most satisfying qualities of our project—simultaneity, surprise, wonder, almost-uncomfortable-not-knowing, extreme open-ness—comprised the texture of our work together. If the project had an overarching sensibility, it was that of falling-in-love—with each other as collaborators, and with carp.

Where do we go from carp?

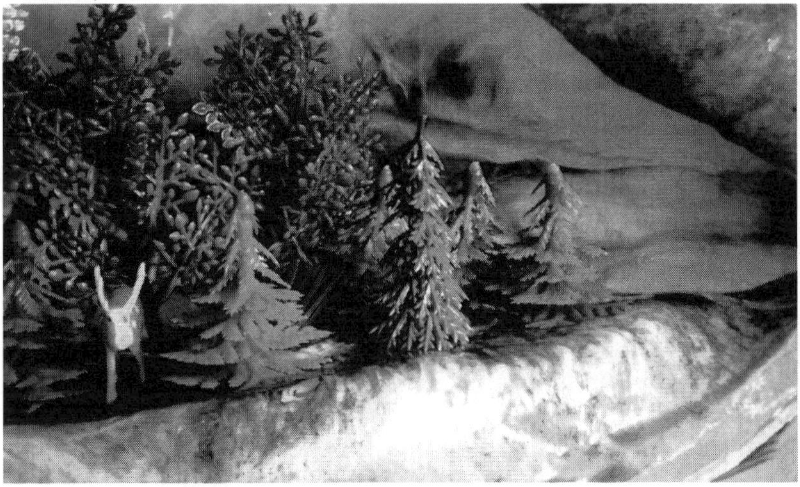

Still from the short film 10,000 Ways to Eat a Carp.

Nature is a construction of human thoughts, emotions, sensations. In an anthropocentric cultural context, nature is everything that falls outside of human making, defined by our need to proscribe the borders of our own identities. When we enforce the borders of these places, we presuppose the authority to shape plants, animals, and landscapes according to an ideal that is often unstated and unexamined. We put a concept of nature on one side of the fence (or electrified dam, or mine pit) and humans on the other.

The stories we are learning about carp, and about invasives across the planet, are part of a broad awakening that's given rise to a new classification of our current epoch, the Anthropocene. The Anthropocene characterizes this geological age as one in which human activity has been the dominant influence on climate and the environment. The shift in perspective this concept represents is on the order of understanding the shape of the earth and its position relative to the sun, or imagining the universe as infinite. A paradigm shift away from anthropocentrism is beginning to produce radically different science as it takes hold, and may rewrite much of what we understand now about ecology and biology, philosophy and ethics. Our existing moral, political, and economic concepts of "empty space," "uninhabited wilderness," and "wild places" will eventually crumble.

In the context of the Anthropocene, natural resource stewardship is even more essential, and yet the gesture takes on an air of absurdity as we currently practice it at the governmental level. I think of another gesture, common in modern choreography, of a dancer grasping the empty air in front of her, jerking and angling forward to grasp more and more. This is what we seem to be doing when we try to meaningfully put our arms around the whole of an ecosystem in order to effectively steward it. Time and again we reach, and time and again we prove to have an incomplete, inadequate grasp of the true extent of Earth's systems as a dam breaks, a species dies out, our planet warms, or our coastal cities lurch toward the sea. In the U.S., we operate with a dangerously slim margin of environmental error, pushing resource exploitation too close to the limits of what we currently know about Earth's resiliency. And it's making us anxious.

At this moment in the Great Lakes watershed, we are putting a lot of this anxiety on carp. We entreat and strong arm carp to help us deal with the exhausting complexity of ecosystems—as well as political and social systems—by being simple and easy. Through incendiary language, killer technologies, budget appropriations, we implore carp, "Just be a fish that wants to fight us, that wants to kill us and what we hold dear. Thrash up out of the roiling river and give us a good clean Hollywood fight, would you please Mr. Asian Carp?" But the story of invasive carp is stubbornly complex.

In our common idea of nature, a few key images stand out: trees, a mountain range, a clear sunset, a coral reef. But mostly our conceptions of nature vary so widely that this is one of our chief problems with our current system: we are not all on the same page about what nature means, or what stewardship ought to mean. All of this is not to say we should abandon our efforts to protect the environment from human-made damage. There are real catastrophes in front of us that our current systems can and must avert. But we must also acknowledge that our stewardship practices today are built to react to a system that relies on unending natural resource extraction.

This is where artists can ask, "What else, if not this?" A question that seems too dangerous, that presupposes the durability of existing efforts, that perhaps ignores the pressure from political and economic systems to do even less than is being done now. But just as carp can't singlehandedly ruin a river, artists can't wreck stewardship by asking difficult questions. Reinforcing narrow systems of value, recreating the same villains in new bodies, doubling down on old assumptions—this is what will harm us. Art can help find ways of moving resource stewardship from reactive to visionary. As artist Jonee Brigham, a friend of the Collaboratory, said at one of our monthly happy hours, "We artists are paradigm workers."

In the Anthropocene, we must move from mass consumption to mass stewardship. The meta-question we asked throughout *The Carptactacle* was this: can we artists help lead a transition from stewardship as a service to a consumerist society, to stewardship as leadership in shaping a resilient, sustainable, equitable society? Even when the stakes are high, we have a responsibility to the future to hold open a space for contradiction and to imagine the unthinkable. This is how we will see new possibilities emerge from the ashes of old systems.

In the Anthropocene our super power is empathy, deep and broken in ways we can't yet imagine. In a program like the Collaboratory, art and science join hands and walk into the scary places of transformation, braver together.

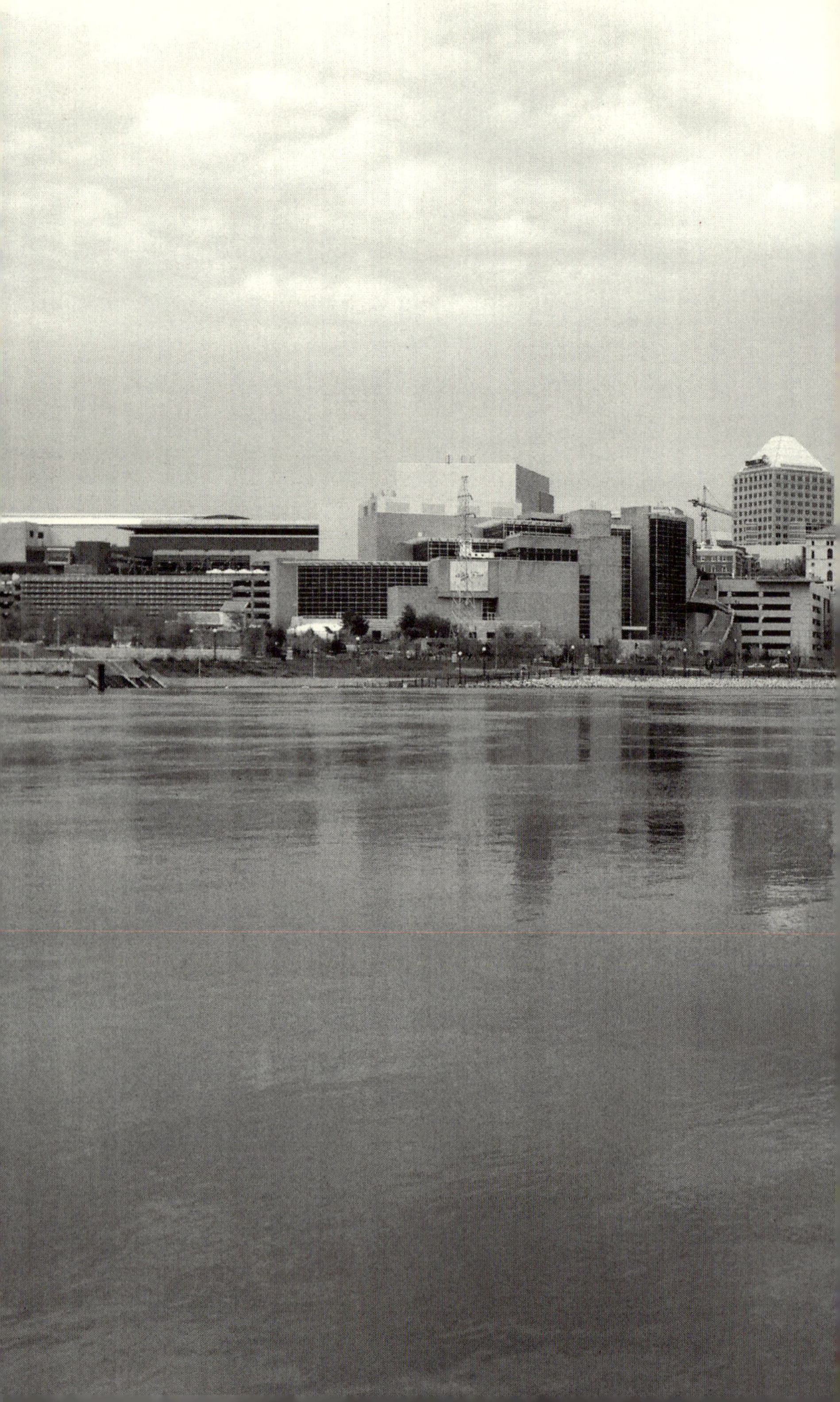

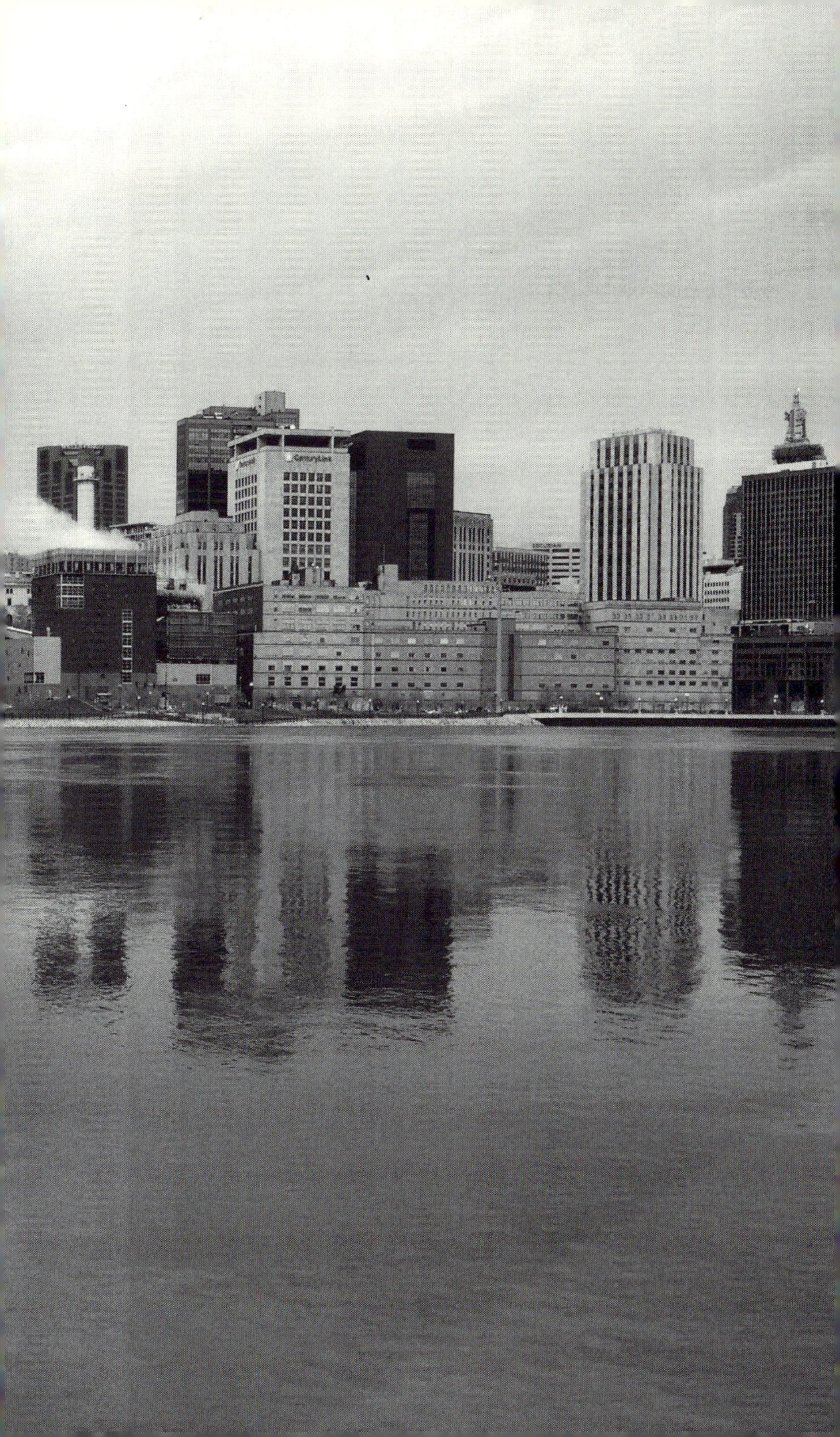

*Preceding pages: Downtown Saint Paul, Minnesota, 2014.
Photo: Shanai Matteson.*

City as Meta-Community: On Climate Change, Resilience, and the Role of Public Art

John Schade & Emily Stover

Ecologist John Schade and artist Emily Stover met through the City Art Collaboratory. Together, they authored this piece about the difficulty of defining resilience, and the possibility of new and different frameworks for urban development.

Urban development is progressing at a breathtaking pace. More than half the human population now lives in cities, the result of an accelerating demographic trend with profound meaning for our social, economic, and political future. The numbers are staggering. By 2050, the global urban population will increase by 2 billion people. In China alone, 40 cities already claim populations over 1 million, and 350 million more people are expected to migrate to urban centers from the countryside by 2025. A new skyscraper will be completed somewhere in China every 5 days for years to come. Shanghai is a symbol of this transformation. The skyline in Shanghai's Pudong district did not exist in 1990; now it is a modern marvel and will soon boast the second tallest building in the world when the Shanghai Tower is completed later this year. Twenty-four million people live in Shanghai, and the population is predicted to rise to more than 50 million by 2050. Shanghai is also symbolic of the challenges presented by our transformation of the Earth's climate. The city is about 15 feet above sea level. If the Greenland ice sheet melts, sea level will rise about 20 feet. Some climate change models project that this will happen by 2100. What will the city of Shanghai look like then?

Sea level rise is just one of a long list of changes we face as the planet warms. Many of these changes are already here, and are creating significant challenges for communities the world over. The western U.S. is locked in a drought that has the Governor of California imposing water restrictions for the first time ever. The Polar Vortex used to be an obscure phenomenon only an atmospheric scientist could wax lyrical about, now it has become a part of our popular culture, credited with the freezing of Atlanta and mild winters in Fairbanks. In 2015, we saw record high temperatures in Alaska and record snowfall in Texas. The ten hottest years have all been since 1998, with

2015 leading the way as the hottest year on record, and 2016 expected to rank even higher. Severe storms appear to be increasing and can devastate large cities that stand in their way. Just ask New Orleans or Joplin or Tuscaloosa. These are extreme examples, yes, but they serve as harbingers of what may come, indeed what climate modelers have been predicting will happen, if we do not take action to slow global warming.

The rise of mega-cities like Shanghai, coupled with the increase in extreme weather as the climate changes, have led to calls for the development of "Cities of Resilience," a new paradigm for urban development that has been institutionalized by new initiatives from The Rockefeller Foundation, the United Nations and myriad other organizations.[1] Resilience has replaced sustainability in some circles as the primary goal of urban design. Building resilient cities has become almost a moral imperative. To many, Cities of Resilience are bright beacons of hope on the horizon.

Resilience has a great deal of metaphorical power. To some, it evokes a sense of stability or strength. They associate resilience with an ability to 'weather the storm' or hold firm in the face of disruption. In the aftermath of natural disasters, survivors speak with

---

1 Mary L. Cadenasso, J. Morgan Grove, Steward T. A. Pickett, "Resilient cities: meaning, models and metaphor for integrating the ecological, socio-economic, and planning realms" in Landscape and Urban Planning, (Vol. 69, 2004), 369-384.

pride of their strength and resilience and determination to rebuild. Resilient people overcome hardship and carry on with their lives. To others, resilience suggests adaptiveness and flexibility, the capacity to bend with the wind and evolve. We are encouraged to raise resilient children because they can adjust or adapt to changes or hardships more easily. A subtle contradiction lives inside of resilience which blurs its meaning and weakens its usefulness as a unifying metaphor to guide the design of cities: resilience can imply stasis or resistance to change, or it can imply flexibility and openness to change.

This contradiction arises from a lack of clarity in the definition of resilience, and a lack of understanding of the conceptual baggage left over from its origins in the science of ecology.[2] We believe that unintended consequences arise from confusion about the meaning of resilience for our perceptions of climate change and the development of policy in response to it.

The rise of resilience as a metaphor for urban design can be traced back to its use as a metaphor for the description of the rhythms and cycles of natural ecosystems captured quite elegantly in the notion of the Adaptive Cycle, a model originally developed by Buzz

2 Kate Driscoll Derickson and Danny MacKinnon, "From resilience to resourcefulness: A critique of resilience policy and activism" in Progress in Human Geography (37(2), 2012), 253-270.

Holling at the University of Florida.[3,4,5] The central idea of the Adaptive Cycle asserts that ecosystems alternate between periods of growth and stability and periods of collapse and reorganization characterized by rapid changes in response to some outside force or disturbance. Resilience in this context is a way of assessing how well an ecosystem resists change or how quickly it returns to its original state or trajectory after disturbance. Scientists studying the Adaptive Cycle in natural ecosystems borrowed resilience as a metaphor to describe how communities of organisms and the flows and cycles of energy and materials would change in response to natural and anthropogenic disturbances. Forests that recover quickly from fire are described as highly resilient. The same goes for fish populations in streams that remain stable in the face of flooding, drought, or overfishing.

Collapse and reorganization leave destruction in their wake, but also open opportunities for change that can result in positive transformation and

3 C. S. Holling, "Understanding the Complexity of Economic, Ecological, and Social Systems" in Ecosystems (Vol. 4, 2001), 390-405.
4 Lance H. Gunderson and C. S. Holling, eds., Panarchy: Understanding Transformations in Human and Natural Systems (Washington D.C.: Island Press, 2001).
5 Craig R. Allen, David G. Angeler, Ahjond S. Garmestani, Lance H. Gunderson, C. S. Holling, "Panarchy: Theory and Application" in the *Nebraska Cooperative Fish & Wildlife Reseach Unit -- Staff Publications*, (Paper 127, 2014).

reinvention. Ecosystem management strategies are often designed to maintain, even encourage, the forces that lead to the destructive periods of the Adaptive Cycle because they play an important role in maintaining healthy ecosystems.

What about us? What does it mean for a human society to go through the collapse and reorganization phases of the Adaptive Cycle? History provides examples that may serve as cautionary tales, like the aftermath of the fall of the Soviet Union or any number of revolutions or riots. The costs can be unacceptably high no matter how much we may anticipate something better rising from the ashes. The image of collapse when applied to human societies is alarming. Change, however, is good. Effective human institutions respond to change with creativity and flexibility. Human societies develop in a tension between the two poles of resilience—the need for stable institutions, and the need for reinvention or transformation. The current emphasis placed on Cities of Resilience highlights the need for clarity in the definition of the term, and the assumptions and biases that drive how it is applied. Not all of the initiatives designed to increase resilience will lead to a civil, just, and sustainable society.[6]

6 Nancy B. Grimm, J. Morgan Grove, Steward T. A. Pickett, and Charles L. Redman, "Integrated approaches to long-term studies of urban ecological systems" in Bioscience (Vol. 50, 2000), 571-584.

The emergence of 'Cities of Resilience' as a metaphor for urban design is a natural consequence of a paradigm of the city as an ecosystem that has taken hold amongst natural and social scientists who have led the way in the development of an Urban Ecology.[7] The 'city as ecosystem' paradigm has led to great progress in understanding how human decision-making interacts with the environmental context of a city to drive flows and cycles of energy and materials. Unfortunately, many in policy and design communities still hold onto an old idea of an ecosystem as a bounded, internally integrated unit that exhibits stability or equilibrium that most ecosystem scientists have discarded. This can result in a strong focus on maintaining stable institutions and a use of the term resilience as a justification for maintaining the status quo, even as we face long-term fundamental changes in climate at global, regional, and local scales. Resilience in the ecological community no longer implies equilibrium. In its more recent form, resilience is defined in terms of dynamics, adaptability, and adjustment to changing conditions. Ecosystems are more like resilient children, than rigid, inflexible but sturdy adults. The behavior of institutions is very likely to hinge on which version of the resilience metaphor lies at their foundation, and will likely influence policy decisions profoundly.

---

7 M. Holyoak, M. A. Leibold, N. Moquet et al., "The metacommunity concept: a framework for multi-scale community ecology" in Ecology Letters (Vol. 7, 2004), 601-613.

Both versions of the 'city as ecosystem' metaphor lead to a perception that external disruptions are immutable. Resilience, in fact, implies that disturbances are a given and are not subject to our control. We must resist or adapt to them, but we must accept them as part of the world we live in. This is true in the context of physical disturbances or natural disasters. But it is misapplied when it comes to disturbances with economic or social forces lying at their foundation. These forces are mutable by the actions of individuals or communities. Does use of the term 'resilience' make this harder by equating anthropogenic forces and physical forces in terms of their immutability?

This is particularly relevant to climate change, which is largely an economic, social, and political problem now. Focusing our efforts on developing resilient communities contributes to a sense of climate change as a process that is beyond our control and reinforces the notion that the status quo must be preserved even at the cost of local communities. The idea of a resilient community as a moral imperative (or as necessary for our survival) serves paradoxically to shift our attention from the need for large scale changes in the social context within which local communities develop, at the expense of local resilience. How can we build a society that catalyzes its own instability, and therefore avoids stagnation and rigidness, but still provides the type of stability that people need to maintain their well-being?

This brings into sharp relief the power of language to frame social and economic life. Metaphors are essential for communicating complex and powerful ideas, but become dangerous when the technical content of words or the conceptual lineage of a term is not communicated. Resilience has a lot of conceptual baggage associated with it, and has been imbued with morality, i.e. has become a path to living in harmony with the world and therefore represents a moral imperative which is self-evident. Co-opting this term to maintain status quo and prevent the type of progressive transformation we often need is counterproductive. Using it without clear definition of the conceptual and intellectual content that it holds can lead to problems.

There is an alternative. Another metaphor has emerged from the ecological sciences that we propose may be a more effective foundation for new ideas about how human societies can balance the desire for stability with the necessity for transformation as the world around us inevitably changes. We argue that cities are not single, bounded, tightly integrated entities as embodied in the 'City as Ecosystem' paradigm. A more accurate vision would consider a city to be a meta-community, defined as a set of local communities that are linked by the movement of people, materials, energy, and information.[8]

8 Ibid.

We assert that the 'City as Meta-community' is a more apt metaphor, and serves as a better ecological context than a resilient community embedded in a bounded ecosystem that is responding to immutable external agents of disruption. A meta-community allows small-scale, local changes to feed back to city-wide features, but also encourages local communities to be dynamic and innovative while preserving stability at the scale of the entire city. A meta-community with a set of shared values can open up possibilities for controlling the collapse and reorganization phases of the Adaptive Cycle. The City as Meta-community comprises many small-scale entities all moving through the phases of the Adaptive Cycle with some degree of independence. This creates a dynamic and creative mix of stable safety nets and agents of chaos, challenging and transforming institutions and worldviews that become stagnant and myopic.

But can we really control the destructive forces driving collapse and reorganization, even at local scales highlighted by the City as Meta-community? What are these agents of chaos and how can we use them appropriately to create change without the consequences of a catastrophic collapse? This brings us, at last, to the role of the public artist.

Art can shake our perspective and create chaos by altering our view of the world, expanding our ability to imagine new pathways of change, or new systems that can work for us. In a sense, art is a means for experimentation with new ways of living in the world, or new types of institutional structures. Artists create folk knowledge that can serve as a framework for creating shared vision based in myths as explanatory frameworks. A rich cultural environment that breeds innovation and creativity promotes change and recognizes the tension between local identity and the larger context of valuing differences. A vision of the City as a Meta-community, combined with a rich and dynamic society of public artists, has great potential to balance stability and change, giving us the power to create the kind of transformation necessary to maintain a civil, just, and sustainable urban society.

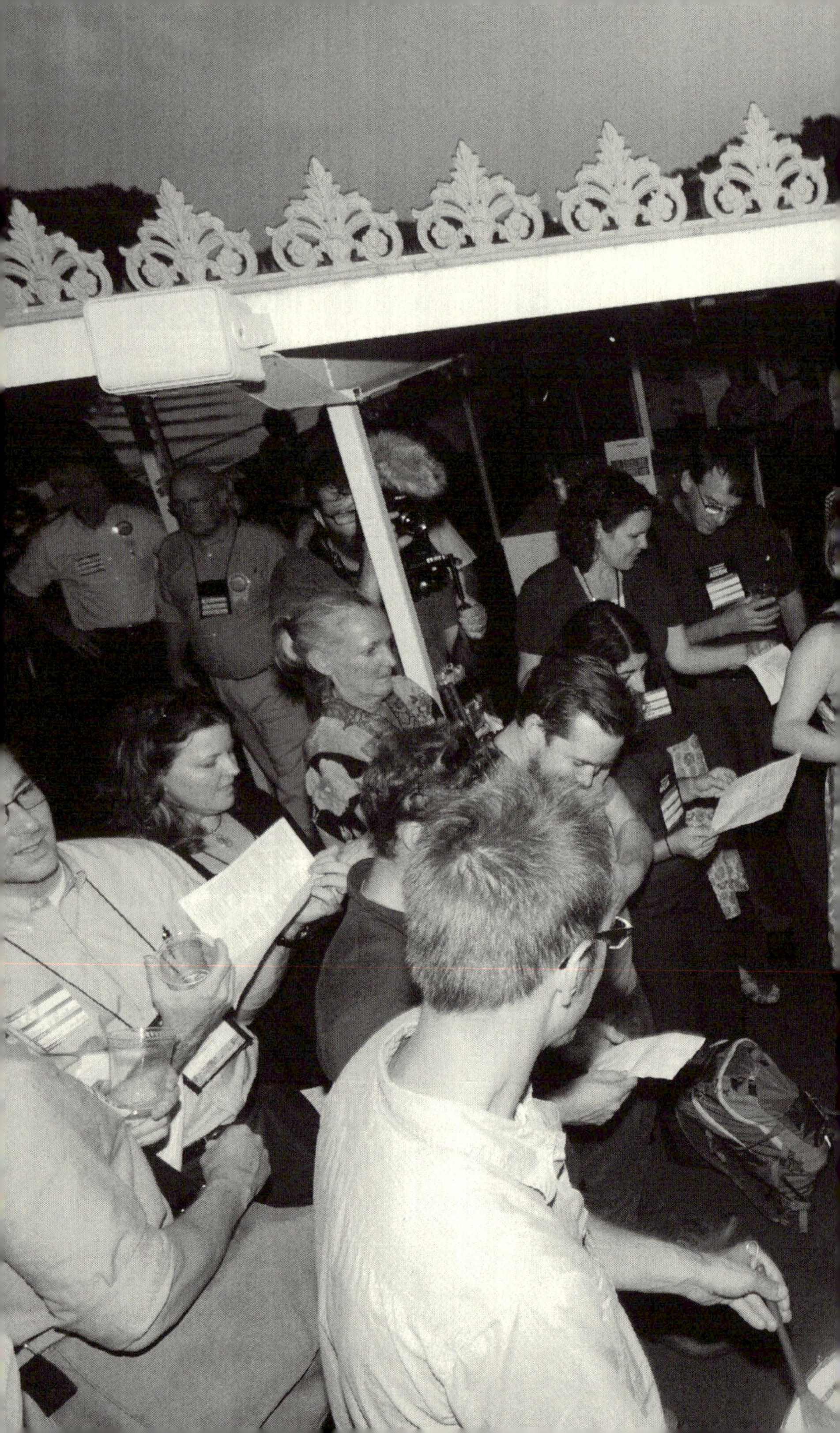

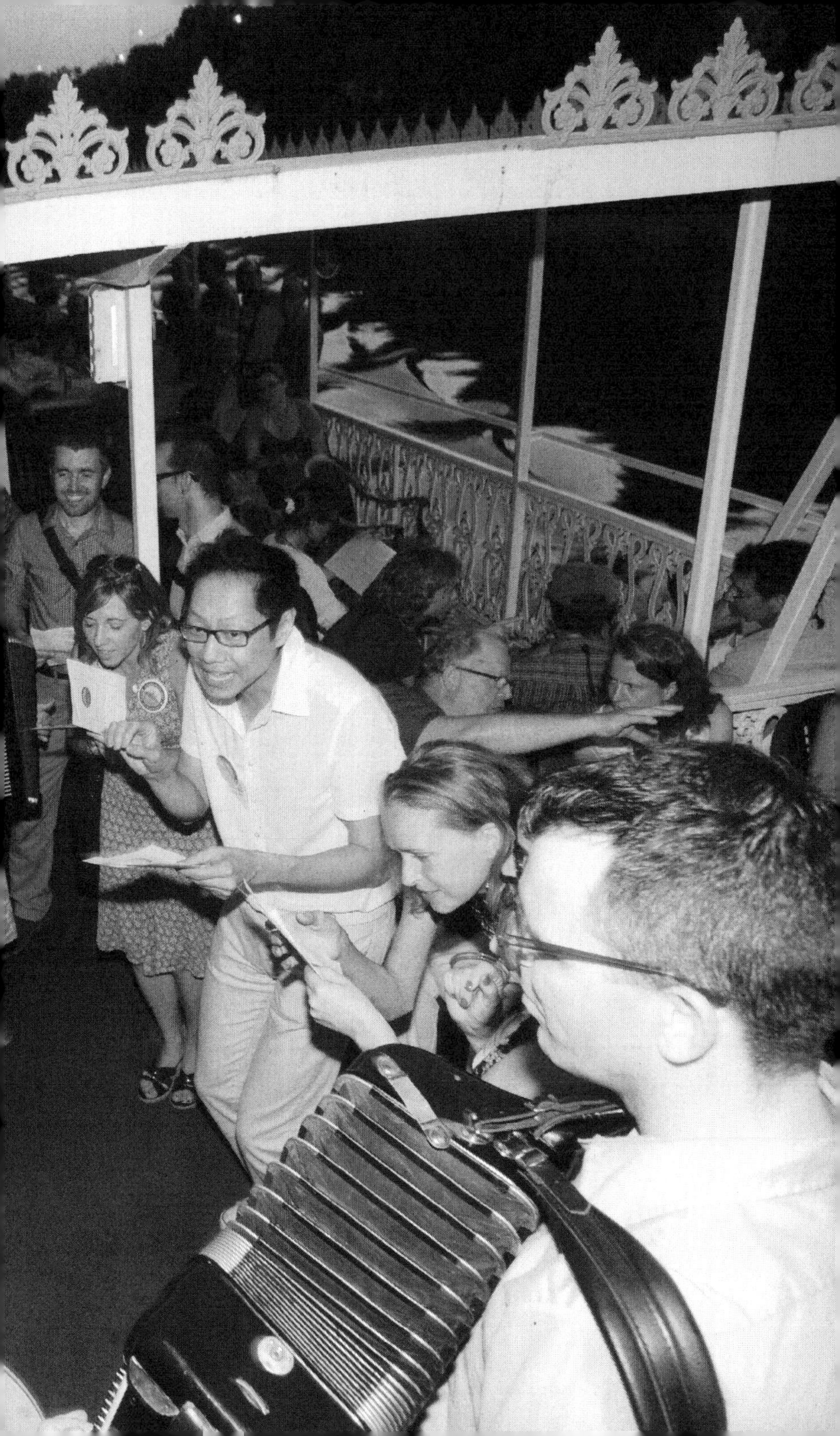

Preceding pages: Collaboratory Curator Marcus Young, Art Fellow Molly Balcom Raleigh, and musical group Dreamland Faces lead a sing-a-long aboard a riverboat chartered by Public Art Saint Paul in conjunction with the LID Symposium in 2013. Photo: Sean Smuda.

Art, Science, Participation, and Passion: A Presentation from the Wild Margins

Janaki Ranpura

The following text was presented by Collaboratory Art Fellow Janaki Ranpura at one of our monthly social events. She speculates on the ways that dialogue between art and science can lead to action.

The terms "art" and "science" have a great latitude of meanings. I will try to simplify some of these meanings just enough to lead us to action.

I'd like to start out by saying that I know art and science have a great number of similarities—specifically, they are ways of knowing the world that seek to open up human experience. Both art and science do that. They also test reality. They both help define our place in the universe.

However, there are also important ways in which art and science are not similar: science proves that the world exists; art disproves the feeling that we are alone in the world. Science observes the world we all live in; art observes the world within each of us. Science seeks to dispel opposite, simultaneous conclusions, while art seeks to hold a dynamism of divergent, simultaneous conclusions.

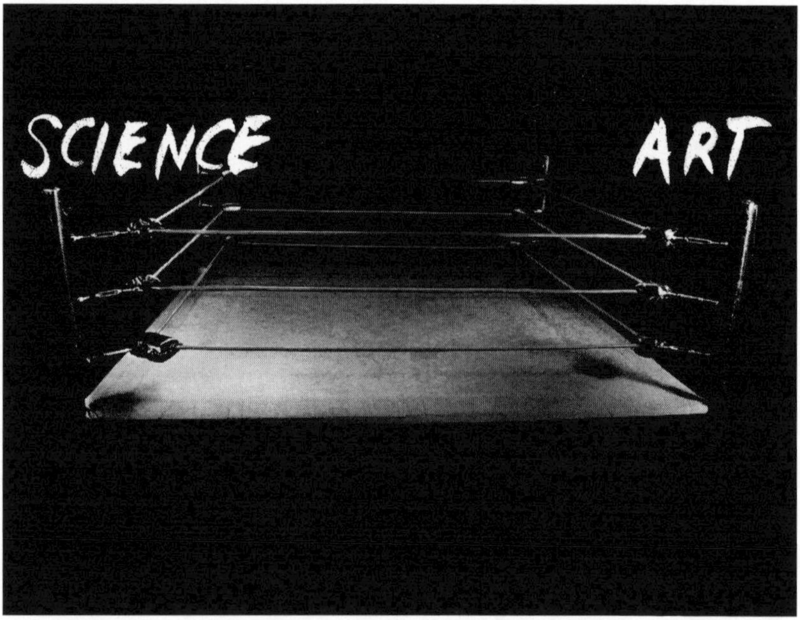

Image from a talk given by Janaki Ranpura at a Collaboratory Social in May 2014.

So, what are we talking about, exactly, when we talk about art and science?

There are some obvious things, grouped into disciplinary categories: painting and stone carving are art, systems engineering and physics are science. But what about public engagement, risk, writing, and action? Many categories overlap.

How do we use all the ways in which art and science overlap to advance toward a shared goal? This is a persistent question of the work of the Collaboratory.

I want to say, as an aside, that I recognize there are problems with these boundaries around art and science, indicating prejudices that art is fine art and science is classical science. Art and science both have theoretical and applied sides, and the means they use to tackle ideas can mush together. This makes it all the more urgent for people who might feel siloed in their fields to have platforms to communicate with people outside of that field.

Given these problems of distinction and communication, I would say that a shared goal of art and science could be simply a dialogue across fields, a dialogue in between art and science that can test its own discoveries, perhaps through objects or products, physical things, or writing.

So let's imagine these objects or products, proofs of good interdisciplinary communication, exist. And they garner some attention, they get noticed for the means from which they evolved. What do you do with that attention?

There are a few possibilities I can think of. One would be to pass this conversation on to politicians, to push it into the realm of policy. To utilize this medium of attention as the non-specialists we all are. And by non-specialists, I mean citizens, or amateur politicians, as we are supposed to be in a democracy.

Another way to use those products or objects would be to influence frameworks. Once those products or objects have garnered some attention or recognition, they can use that attention for news about them -- or, the lessons embedded in them -- to spread. This can look like marketing, or it can even look like propaganda—and maybe it is propaganda.

When we as artists and scientists get into the process of communicating what we think art and science are achieving, all kinds of boundaries quickly become blurred.

This interdisciplinary space is to me exactly the most exciting space. It contains the reality of many constantly changing viewpoints. After all, within a forest, 90% of the life in that forest exists along the edges.

So, in that edge space, teeming with possibilities, how do you tell where to turn your attention? This is where passion comes in.

By passion, I mean what takes up your attention. I also mean to include the effect of passion: by taking up your attention so completely, what you are noticing attracts the attention of others.

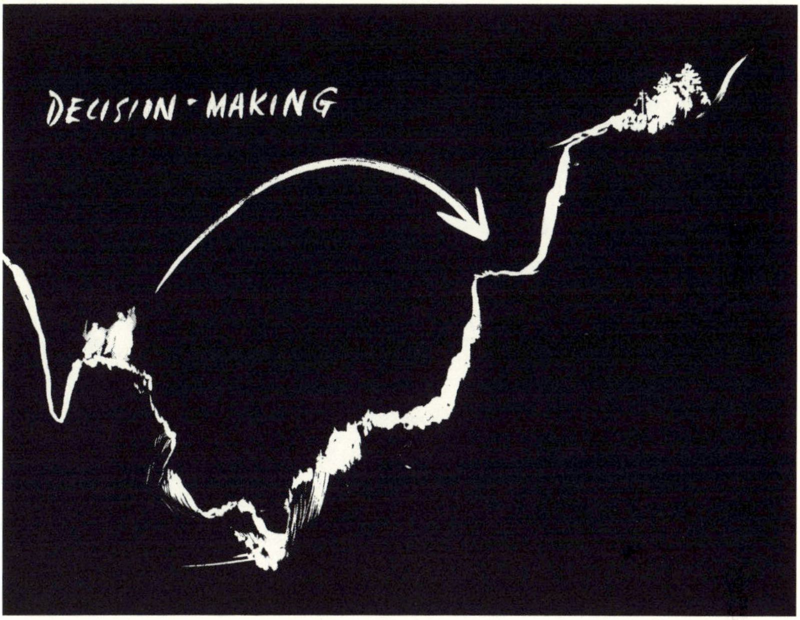

Image from a talk given by Janaki Ranpura at a Collaboratory Social in May 2014.

Passion makes us stick to things. It answers, "Why am I doing this?" by saying, "Because doing this is living." Passion is very influential in decision making. It can, however, be hard to accept as a rationale.

We often operate under the idea that we can accumulate evidence and make decisions based on that evidence. But there is a gap in our process between thinking rationally about a decision and making a decision. Our rational faculties churn away until they get tired, then we simply leap to a conclusion.

What about that leap? It's that gap between the end of rational thought and coming to a decision where education, knowledge, passion, and prejudice all operate. The gap of decision-making is the most important point of influence. Two books in particular have led me to these thoughts: *Predictably Irrational*, by Dan Ariely, and, *Thinking Fast and Slow*, by Daniel Kahneman.

So, there are formal ways of educating people about the rational part of decision-making, the part we think is influenced by a logical sequence of thoughts. But how do you influence the irrational gap? How do you influence it often enough to tell you're actually influencing it?

Making information accessible to many points of the social spectrum is important in order to build consensus. How do you influence that irrational gap in people across a really diverse spectrum of rational inputs—meaning, how do you influence that irrational gap amongst people of very different educations and upbringings?

Here we run into another problem, equally shared in art as well as science: results are often considered less valuable if they only affect a group of like-minded people. And results that reach a truly heterogeneous group of people risk being dumbed-down, artificial, or just impossible.

However, assuming a way forward is possible across a broad spectrum, I speculate that that way is by communicating passion. How do you communicate passion? Honestly, I don't know. Silly but technically obsessive drawings, sculptures the size of hillsides, terabytes of immaculate data, infinitely durable and intelligent precision machines, free beer?

Something somewhere in the wild margins.

Communication via language is one of the most sophisticated things that we do. It actually helps us get a little further with each generation, because when it's used and received well, it can help us avoid mistakes by proxy. The better we are at developing a form of language that communicates the specific things we know, the more space we open up for change in our species—meaning that you can learn from the results of a life you haven't had to live, and mistakes you haven't had to make, and even benefit from those mistakes, simply by communicating via language.

Opening up space for change in our species is not a measurable outcome.

Outcomes are really the place of policy. I'd prefer to think about directions. Impact is also a nice word, because it indicates a crash that can cause a change of course.

So I would ask, are the people framing inquiries—the ones looking and researching, the artists and scientists—the best ones to evaluate the outcomes of their questions? Or does that actually require a fundamentally different way of thinking, a different language, meaning a different perch from which to view the world?

I would propose that a list of markers for a good project in an art and science collaboration would include the following:

- What framework is being influenced? Give it a name and qualities.

- Where is the passion / danger—what can and should be risked for the sake of influencing this thing that we as a group have named?

- Does the project use science's ability to prove something about the world in which we live, and art's ability to dispute that we are each alone? In other words, is it factually sound? Intellectually interesting? Novel? Does it have an emotional impact that helps us confront our individuality and help us feel we are not alone?

- Does it cross into policy? Do we want it to? Do we have the right tools and people in place to have a responsibility for impacts? At what point are we quit of the spread of impacts—in other words, for

> how long are we committed to stewardship in order to see this project done?

I am eager to see how wily we can be at influencing each others' passions, and how malleable we can let ourselves be.

What else stands in the way of action?

An audio slideshow of Ranpura's presentation can be found online at janakiranpura.com/wildmargins.

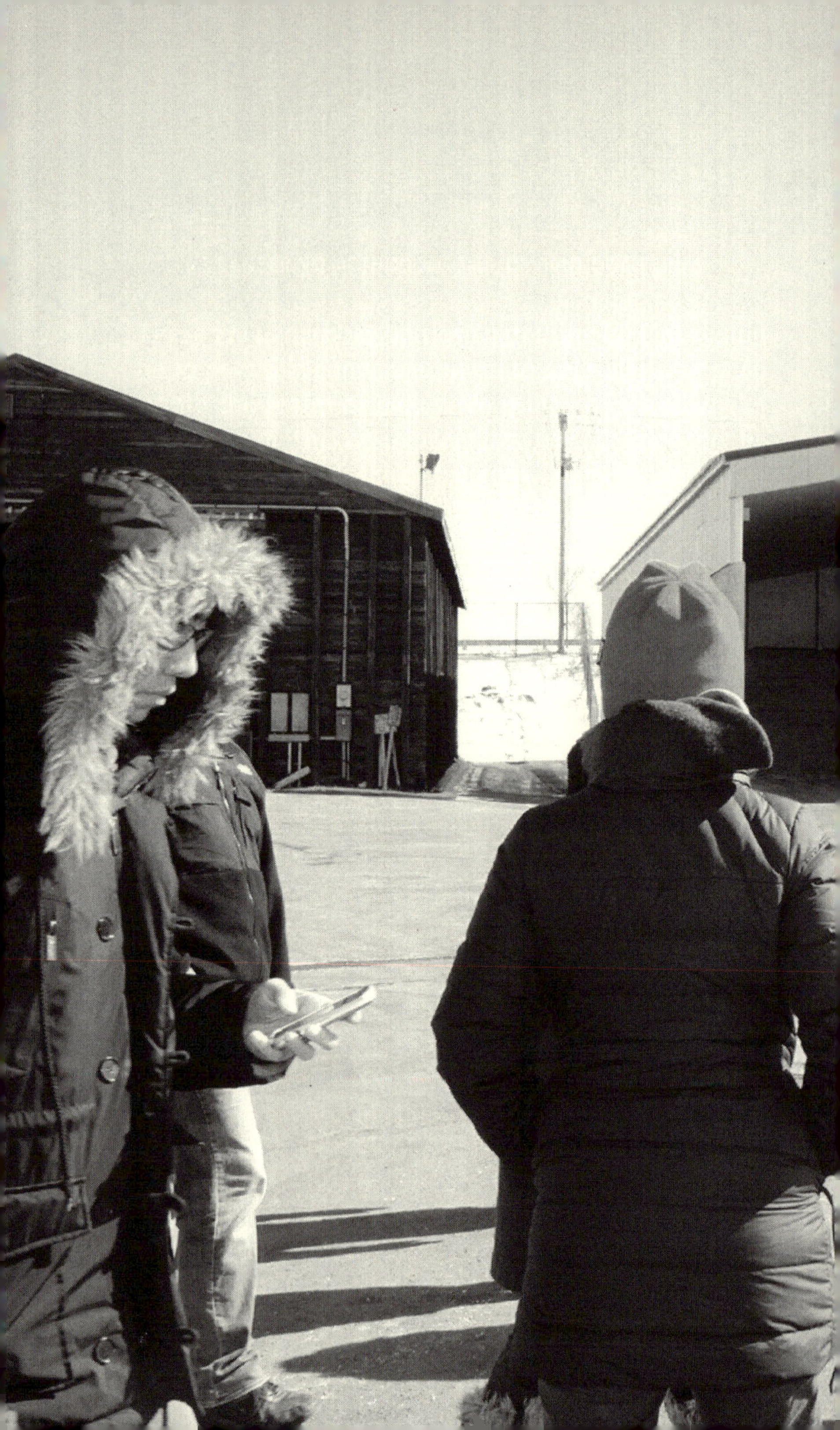

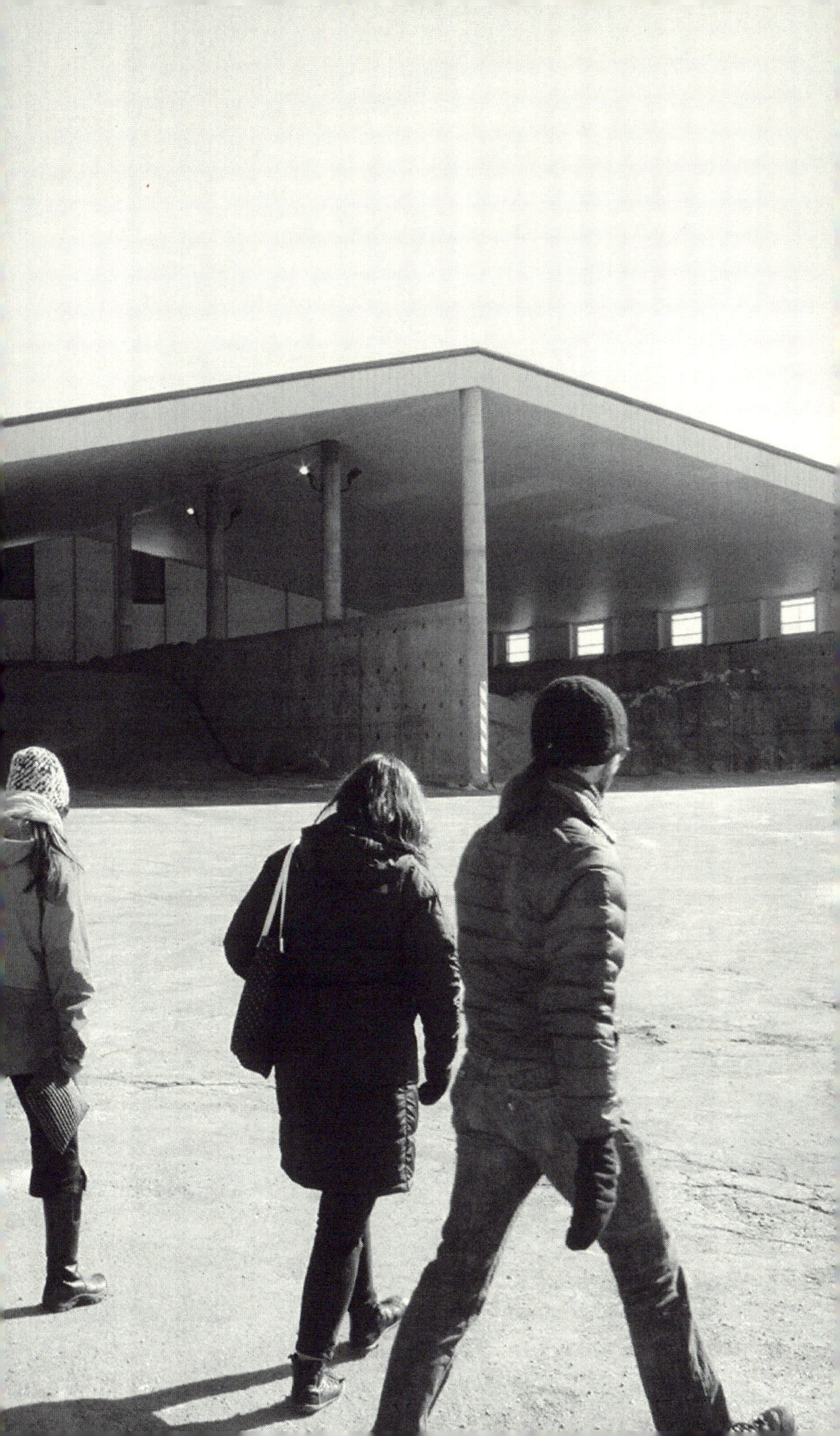

Preceding pages: From a field trip to the City of Saint Paul's Street Maintenance Division in 2013. Photo: Andrea Steudel.

Methodologism: Radical Practice

Kirsten Valentine Cadieux

As a Collaboratory Science Fellow, Kirsten Valentine Cadieux continued her exploration of the social and spatial relationships related to landscape and food production. In this piece, Cadieux reflects on what arose for her during the Collaboratory project, including questions of methodology, platforms, and elders.

Methodologists on the Verge[1]

We are about questions.
We are compelled to answer in as many ways possible,
because we know our starting place is not quite right
and we have not assembled the authority to make
  what we feel
we need.

We are pushing beyond the end of history. Another

1 Written for the *Art(ists) on the Verge 6* catalogue, from the exhibition at The Soap Factory, Minneapolis, MN, 2015.

world may be possible,
but we have to find where it is already made
and figure out how to fit our hands and voices in
because if we are successful, we risk being louder than the silence we heard
when we knew
no one was listening.

We know we must practice, and we make our ideas
for others, so they can know the world they make too.

We place people:
in our violence,
in our neuroses
and our lusts,
in wanting to be known,
and to have the authority to place ourselves on the maps of what we see matters.

We hope methodology may bring us new access
to experiences and elders,
as we plumb beyond narcissism
how broken things are, and hack the systems we live within
using the healing alchemy of process to find delight sharing methods.

We are our own critics,
not because we have authority
but because the whole point of methodology
is that we put the practice into the circle of those we

want to care,
and see what we can do with it.

We Are Our Own Critics

For the past four years, my participation with the City Arts Collaboratory has provided a social space for organizing collaborative explorations of a number of practices I have often wished I had the social support to prioritize more radically. Focusing particularly on popular experiences of environments along the Mississippi River, and on city systems that artists and scientists might help people experience more vividly, we have spent many weekends figuring out what kinds of practices we can share to make the world we live in more comprehensible, meaningful, and collaborative.

As we reflect on this collaborative experiment, I have gathered a bundle of the themes that surfaced repeatedly for me on our field trips around Saint Paul, and in the many Collaboratory projects I participated in, almost all of which were organized around art practices designed to make the process of interpreting meaning popularly accessible. Although I realize many of these ideas might be better experienced in practice, I here sketch a trajectory of thought through some of the struggles to support and institutionalize the kind of social practice that has been the water in which the Collaboratory has been swimming, focusing particularly on the value of our relationships with methodologism, platforms, and elders as important parts of radical practice.

Those of us who engage in social practice art and science are often asked to describe it—our emphasis on process, and indeed practice, can come across to people accustomed to engaging with art via the observation of objects or performances as ephemeral, abstract, or confusingly un-art-like. And our engagement with the phenomena of experience and the meaning-making process may seem avant-garde for science. In response, explanations of how, in fact, social practice can excel at distilling key parts of what makes art important have become central to this field, and its corollary is starting to appear in science, especially around environmental social practice. Part of what seems striking about the emerging social practice canon is the way it is narrated by practitioners, rather than by external reviewers such as critics. One feature of social practice this may reveal is the breaching of traditional barriers—often barriers of class, privilege, race, gender, and different modes of communicative fluency—in art practice and interpretation. This communicative barrier has been shaped differently in science, and may have more to do with what scientists are afforded in both scientific and humanistic training, itself a question of privilege given habitual segregation in access to different modes of education.

In exploring these ideas, I do not mean to sidestep important questions around the reproduction of power and privilege in art, and likewise in science: granting agents, institutions, collectors, and canonical establishment voices have tremendous influence over the lives of practitioners—as, increasingly, do municipal and natural resource bureaucracies as well as other placemaking and social institutions that are engaging artists and scientists (sometimes together) as part of multifunctional approaches to the complex challenges facing us. I do, however, observe that some effects of the incorporation of social practice art within such institutions involve a transformation of who perceives themselves to participate with agency in art and science practice—and who has the authority to discuss, describe, and interpret it. This transformation has implications that may be useful to note more explicitly, and to consider as part of the field of possibility when we explore new projects and ongoing challenges.

Introducing the three main artifacts I consider here—methodologism, platforms, and elders—I note that these are not an exclusive list, but they have helped me orient myself to the tasks of the common project in the often meandering processes of collaborative practice. I consider methodologism a crucial stance for shared practice. It provides an explicit focus on "how?" questions, while keeping methods, methodological justifications, and integrated evaluation of those methods, strategies, and tactics constantly in play. The platforms that hold the space for the methods in question may be the material and conceptual toolkits of methodologism. For example, platforms can help delineate the modest scope of a particular exploration, such as the attempt that some of us Collaboratorians made to explore and articulate the value of elders in social practice work. The value and practice of elders can obviously not be contained in an essay, or, arguably, in such a limited exploration as ours, but the platform of the thought experiment "we are our own elders" pushed us to distinguish between the relationships that compel us to produce meaning in various ways, and those that call and teach and sustain us.

Our search for our own elders arose out of frustration with modes of work that exploit our practice by promising public exposure, participation in institutional contexts, and proximity to famous people instead of meeting our needs; with modes of analysis that see these proxy rewards as markers for a successful climb up a dominance hierarchy many of us do not wish to participate in; and with the still-dominant kind of teaching that tells you how it is, from that teacher's perspective, and how it works really well for them, in a way that neither invites you into a path to discover for yourself, nor foregrounds the likelihood that you will travel some parallel and well waymarked path. Figuring out what we needed from elders, and how we could be in relation to them, involved both seeking them out on our Collaboratory missions, and also playing the role of elder to each other. We discovered things that may be so obvious to people who have had the privilege of elder relationships but that are radically novel enough for those who have not that they may be worth saying out loud. For example, one of the things with elders that's different than relationships most of us have had with various teachers is that elders agree in some way to take responsibility for you, and you agree to consult—not like an employee or a child, maybe more like doing advanced study in an area where you will be in a lineage and the trainer wants to make sure their knowledge is represented adequately. The idea of negotiated reciprocity in a relationship with elders added more practices to our methodologism. These practices for relational accountability were especially useful in decolonizing struggles. Struggles with forms of domination and hierarchy characterize the chronic stress that prompts the need for the kinds of transformative work the Collaboratory fomented. By also prompting our search for elders, struggling with these stresses also provided helpful vocabulary and practice for facing the ever-present challenge in our settler context, of people, especially white people, ever knowing who they are themselves.

Platforms are designed to foreground the process and approaches used in relation to a particular path. And platforms can have a similar role to elders: they remind you that part of the work is attention to how you're doing the work. From this perspective, documentation and sharing aren't just by-products or necessary evils imposed by grantors or our institutional contexts. What is produced in the method—the path—might be just as important as the destination.

This is a key frame of much social practice art and science and the kind of social organizing of which it's often a part. This attention to method can help constrain the dominance of people accustomed to taking privileged roles within a community trying to collaborate. Without showing acknowledgment or awareness that their knowledge might be alienating or annoying, such assertion of an agenda or method can be incredibly corrosive to collaboration. Explicit and methodical engagement with the methodologist approach to platforms, and to the reciprocal relations with elders can help call in a radically contrasting mode of action. Instead of passing along knowledge and authority, those vested with power in this model support and invite you into action.

An elder, as we understand the role, may be the person you might run an idea by, and whose restraint or approval you might be looking for. Along those lines, elders might also be expected to have a sense of your capabilities—or when you might be ready to do something. In this relationship, there might also be the possibility of them not just encouraging you to do things, but helping you figure out what the signposts of success are, or how to get feedback about what's working or not. We thought about the possibility that an elder relationship might be one where someone is encouraging and exploring along with you, not just encouraging you to go where you've already gone, but helping to find and set some goals, guardrails, and way markers. Elders might help to figure out what the path is, and to keep figuring out what the path is, maybe in a particular community engagement or project—or stage of life.

We recognize the limited scope in our exploration of elders. We are ourselves novices at this elder question, and in addition to what we are leaving out, we may be romanticizing ideal elderdom. We acknowledge the possibility for the abuse of the status that comes with age. We nonetheless value relationships with elders in part because of the way they address the recurring challenge of helping people to recognize their privilege and not dominate people with it. This might be the kind of relationship that can be cultivated through a course of study, or performed as a service for one another. We really liked thinking about how to be in this kind of relationship, and how to live up to it, and how this relational accountability is part of what methodologism might enable.

In these ways, the reciprocal elder relations are helpful in non-hierarchical community building. The platforms are helpful for inviting people into practice. By sharing practice experiences, platforms can help make clear what kinds of practices are being used. They can help orient people to how to engage these practices, without assuming they know how to already. This can avert the stressful feeling of needing social contracts for social practice that implicitly expect participants to laboriously accrue every social engagement tool they have encountered along their paths. Instead, a platform allows people to put tools down and pick them up gracefully again, because tools and instructions about how they might be used will be what's on the platform.

Methodologism in Practice: We Do Not Have to be Our Own Elders

A few concrete examples of this work may help ground these reflections, especially for readers who have not experienced something like the Collaboratory's working method. Projects with Molly Balcom Raleigh and Janaki Ranpura stand out for me as highlights of our thought experiment along the lines of being "our own elders." Inspired by the post-normal science idea of "extended peer review," over the middle year of our involvement with the Collaboratory, we played with a series of metadata projects, developing prompts and criteria for collaborators to use to share methodological notes about their working processes, and to help co-curate each others' work. We also created more and less formal mentorship agreements, attempting to build supports for exploring this quest for elders, particularly around Molly's participation in the *Art(ists) on the Verge 8* program with me as her mentor—and her as my mentor on mentorship.

Given the challenges involved in the question of how to collect, document, and exhibit social practices, it was stimulating to engage each other not only as a community of practitioners, but also in swapping roles and assigning each other (and others we recruited as part of an extended group of peers) critical, analytical, and curation tasks, engaging those roles, too, as being part of social practice. The shift in authority involved in this reciprocal role play leaves considerably more space for the legitimacy and value of self-interpretation. And especially in the interactions with the culture of science, this role exchange sometimes revealed ways in which social practice art may reward more verbal fluency than some other art forms, or perhaps privilege non-verbal fluency less, making it particularly valuable to be able to step in and out of critic and curator roles in relatively low-stakes settings and back into one's project with new views. The reflexivity generated in this role-shifting and -sharing also highlighted the proximity of social practice art to much social science, and to humanistic scholarship such as critical media studies, with movement toward considering effects, audience, and power dynamics, simultaneously reaching into very abstract domains and operationalizing these abstract explorations in very popular domains through public platforms.

All of this practicing—making the methods something to be tried out, practiced, changed, tried again, rather than something to be judged on for an end result—ended up making us at least somewhat less vulnerable to the institutional power relations we had been so stressed out about when we decided we needed this exploration. Learning what to look for in elders also led us to learn a bit better how to present ourselves, both to elders and to everybody else. Learning that we could be elders for each other, if needed, let us see the poverty of only relying on ourselves, and let us relax more into the everyday being-related that is necessary to actually live with the elders we may already know. I do not know if there is a word more specific than "relationship" that describes the sense I have of what we gained—something like "self-possession," but more collaborative, less possessed than "inhabited." This must be what it feels like to have been in residence together.

We Are Our Own Elders

We, trying to figure out how we
relate to the middle passage,
are not unable to hear
the stories meant for us;
many things you describe
must be rediscovered,
and these are not our fears.

Care, calm, fury, ease,
decades of retold learning;
litanies of cautions,
prcparation, finance populism.

Can we admit to being lost
on who to ask who we are?
We contest equity being led
by derivation and securities.

Calming logistics traffic
requires us to link elbows
and sing, dancing around fires
we teach each other to tend.
The joyous abundance work of the kitchen
is what breaks the curse;
our reading the recipes together
calls us to decolonize.

Since we cannot know who
we are with responding to so many demands,
it is hard to know who can tell us
who we are together.

We crave teaching and appreciate
invitations to disciplines of practice.
We take direction from what ancestors
tell and the beloved land.
Stories we tell into being for these times
share uncertainties we strain to hear about.

Possible elders, ancestors,
how do we know who you are?
Beyond the stories told to shush
our extra questions after bedtime,
we hear you in what we tell each other.

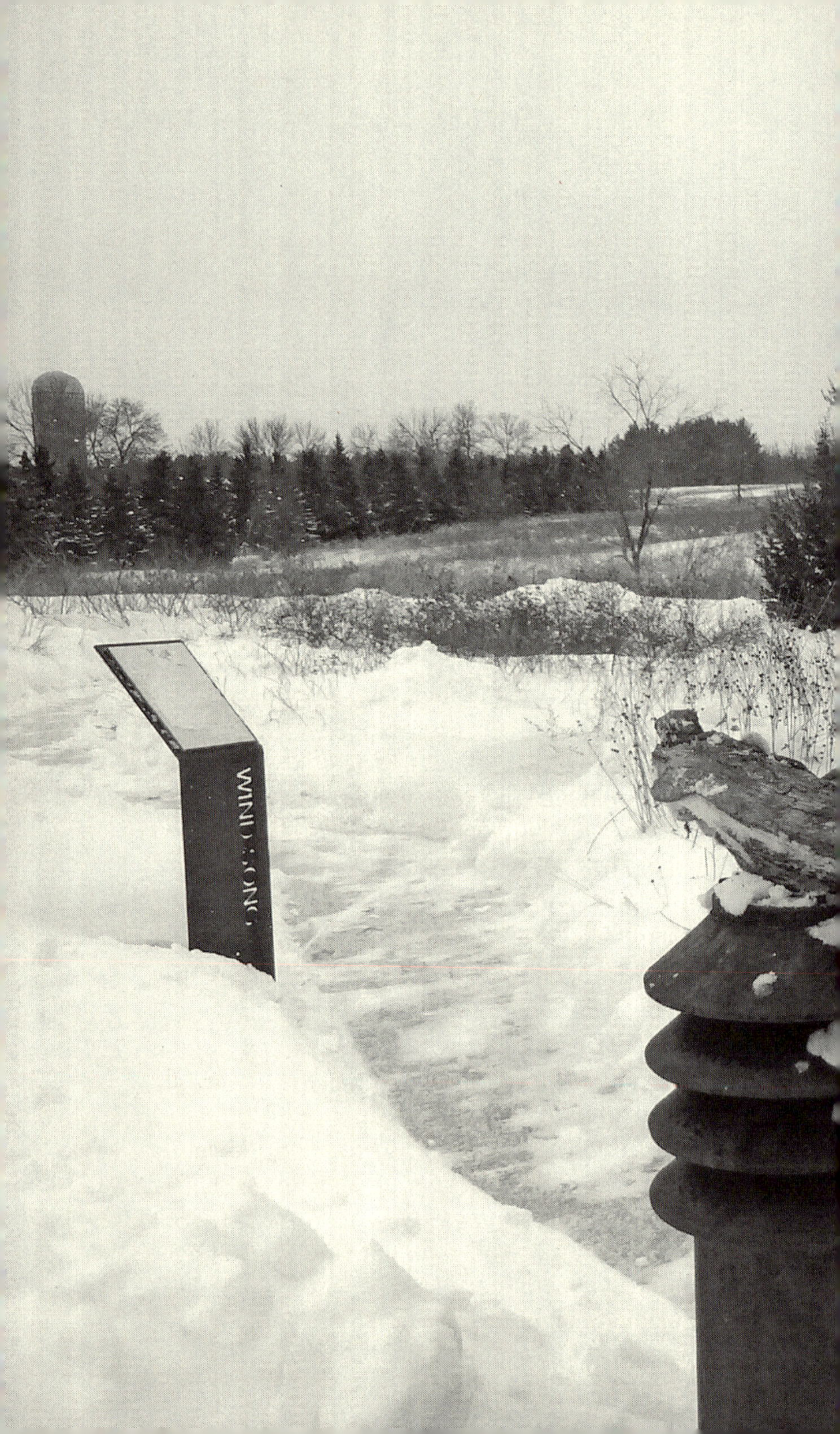
WIND SONG

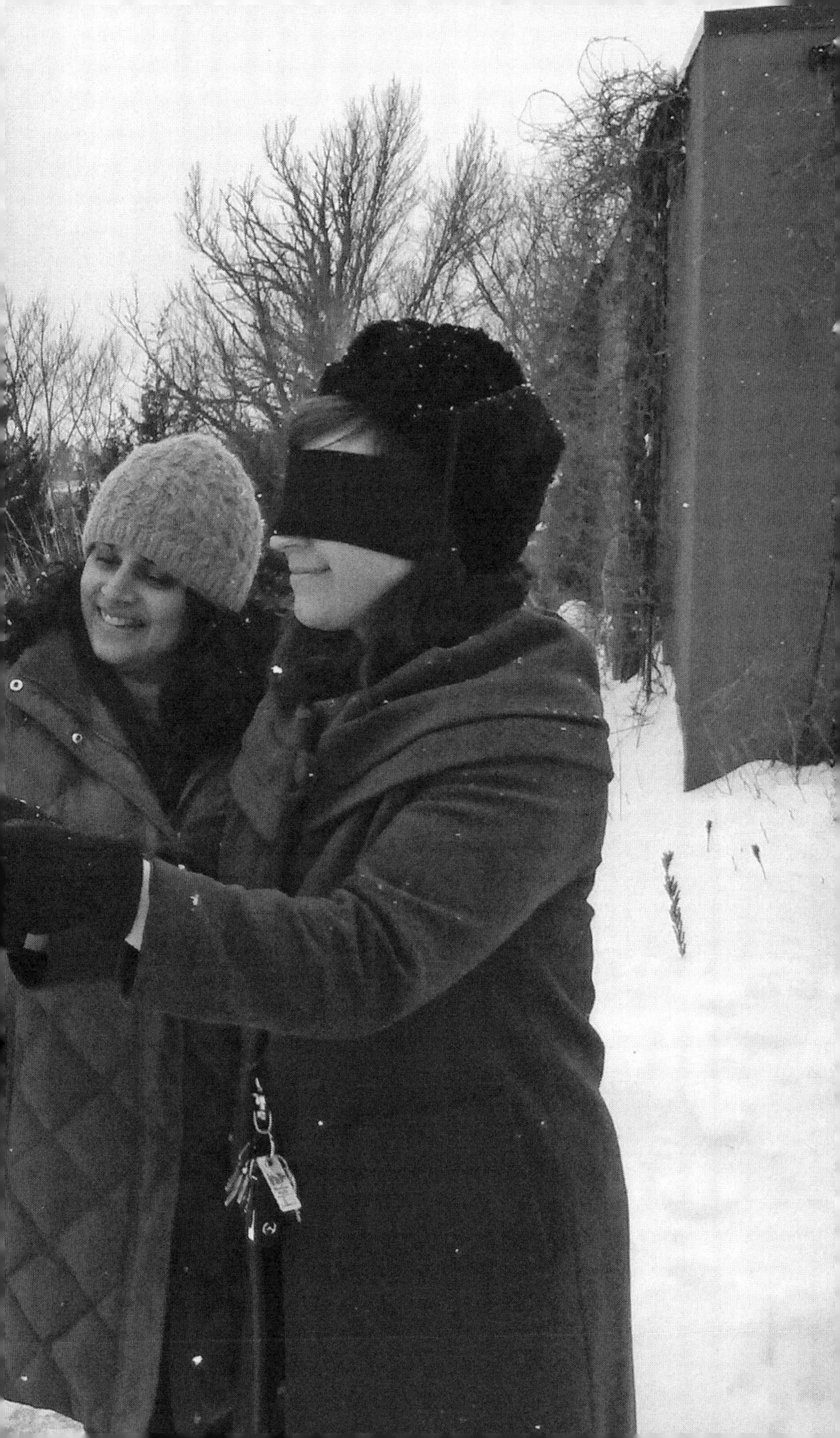

Preceding pages: Collaboratory Art Fellow Molly Balcom Raleigh and Science Fellow Roopali Phadke on a Collaboratory retreat to Schaar's Bluff Gathering Center. Photo: Andrea Steudel.

Deviant Disciplines

Shanai Matteson

Four years ago, when Public Art Saint Paul launched the City Art Collaboratory, our priority was to support the capacity of artists and scientists to work collaboratively across disciplines, specifically those interested in environmental questions or causes. Another commitment we made early on was to embrace an emergent process, following Collaboratory participants in their evolving efforts to explore, communicate, and even transform relationships between people and environment. We had much to learn.

This initial phase of the City Art Collaboratory—documented throughout this book by organizers, participants, and outside critics—brought a cohort of visual, social practice, and performance artists together with research scientists, engineers, and environmental educators. Over the course of several years, Collaboratory Fellows immersed themselves within city systems, applying their varied perspectives and methodologies to the task of imagining what the future of collaborative creative practice might mean for the sustainability and resiliency of people, places, and ecological systems.

Recently, in a conversation with Collaboratory Fellow Kirsten Valentine Cadieux, I realized I'd momentarily forgotten whether she'd started in the Collaboratory as an artist or scientist. Today, Cadieux is involved in many projects with other artists and scientists from within and outside of the Collaboratory. Like others in the Collaboratory cohort, some of Cadieux's projects manifest as art, while others manifest as social science research, public education, or new public engagement platforms.

Does it matter how we name or locate this kind of creative practice, which doesn't just cross disciplines, but deviates from defining them? After four years spent in conversation with the artists, scientists, and others of the Collaboratory cohort, it has become increasingly difficult to recognize what is art and what is science, or for that matter, what is neither or both.

One unexpected result of the Collaboratory program has been the joyful blurring of these disciplinary boundaries that served as our initial organizing framework. While there are important differences between the work of artists and scientists, especially the cultures surrounding art and science as professional practices, many of our Collaboratory activities and conversations circled around similarities. We spent a lot of time articulating a need for alternative ways of understanding and practicing both. In other words, rather than prove what artists or scientists contribute to existing processes, we asked how collectivity and collaboration make it possible to resist the further entrenchment of people within disciplines, as well as the unspoken hierarchies often at play. What can a more holistic and deviant approach to creativity and culture-making contribute to changing the ways urban life is understood and sustained?

Collaboratory Science Fellow Adam Kokotovich introduced our group to the phrase "deviant interdisciplinarity," a fitting term for what evolved from the City Art Collaboratory over the past four years.[1] Deviant interdisciplinarity is distinct from what some call "normal interdisciplinary" in at least one important way. Normal interdisciplinarity prioritizes the organizing of inquiry and practice into distinct disciplines, each with qualifying education requirements or specialized knowledge. It asserts that the correct mix of each discipline will lead to a more productive outcome. Deviant interdisciplinarity, by contrast, sees these silos themselves as barriers, and questions the underlying hierarchies that determine whose perspective or knowledge is valued in a given process.

Problems can arise when the boundaries between disciplines remain unexplored and unchallenged. It is precisely the kind of thinking that leads to declarations

1 Steve Fuller, "Deviant interdisciplinarity as philosophical practice: prolegomena to deep intellectual history" in *Synthese* (Netherlands: Springer International Press, vol. 190, issue 11, 2013), 1899-1916.

like, "We must have an artist at the table!" but which can't bear the question of what an artist is for. In these cases, who benefits most from having artists (or any other creative thinker) at a table that has already been set with pre-determined expectations and outcomes?

Deviant interdisciplinarity proposes a catalytic mixing of methods, and a democratizing of knowledge and authority. It promotes open-ended inquiry, recognizing that there are different areas of expertise, and different experience, but resists the urge to be reductive about it. While normal interdisciplinarity asks us to merely represent our disciplines in a process that is already mapped, deviant interdisciplinarity asks us to question them, and to proceed from new combinations and assemblages.

I've come to believe that by moving ideas between a group of people who see and work in these deviant ways—and introducing those people to different parts of the social and ecological systems of our city—we increase the capacity of individuals to compare, contrast, and ultimately deepen their understandings of themselves and the places they inhabit. They also get a more complete and nuanced picture of the systems on which our urban lives depend, and an awareness of various leverage points for making positive change.

While modest in scope, City Art Collaboratory was a rare opportunity and a valuable model for artists and scientists to learn directly from one another. Together, participants developed a greater understanding of their respective roles and practices, as well as how to move in new and challenging directions.

Many artists today care deeply about environmental sustainability and the future of our planet and are looking for ways to utilize their skills and energy for the betterment of the places where they live and work. In the fields of environmental research and engagement, new ways of working with and in public are vital. I believe we can continue to connect these efforts in ways that are truly transformative—beginning with an exploration of who we are, where we are, and what is possible together.

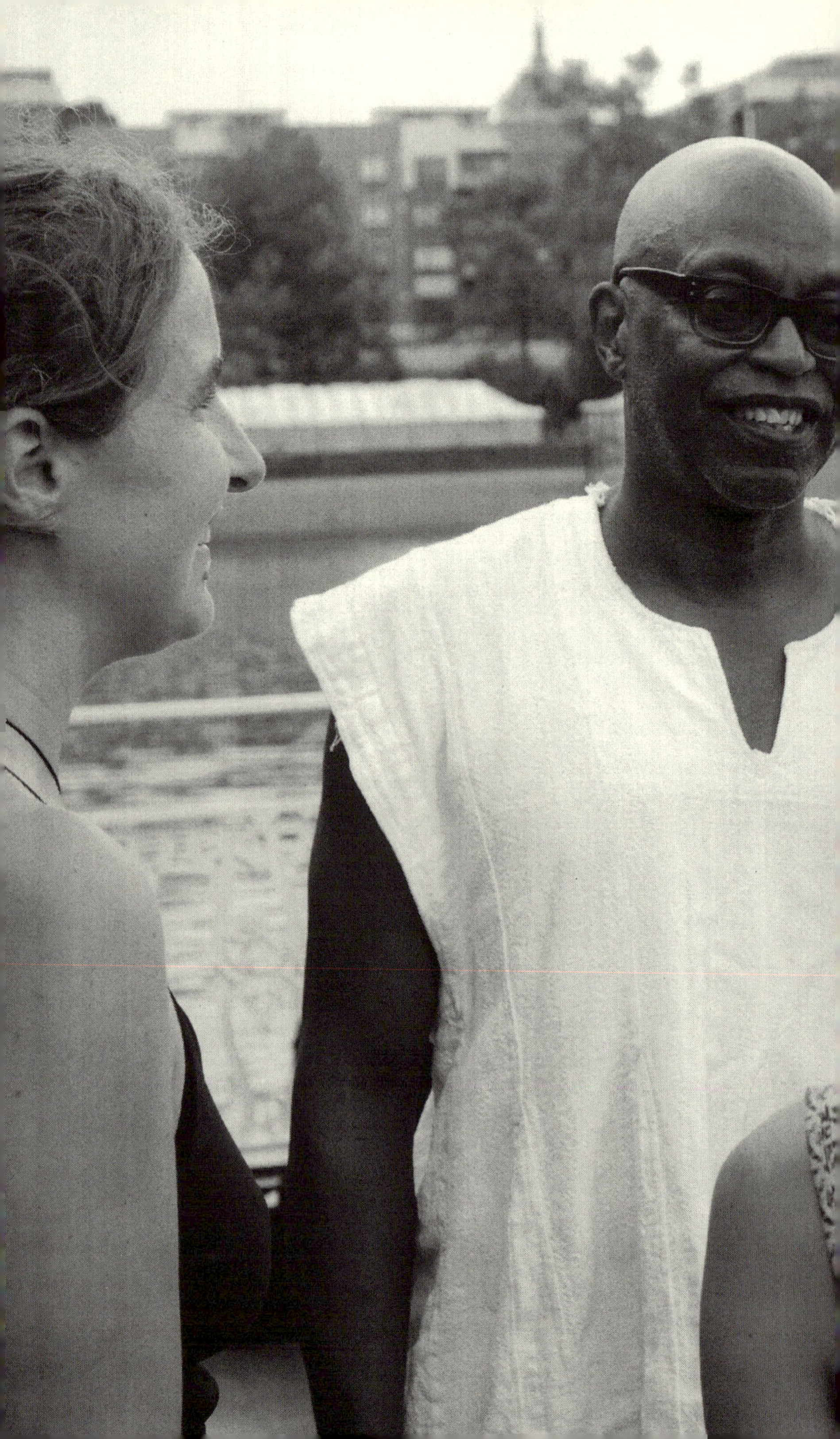

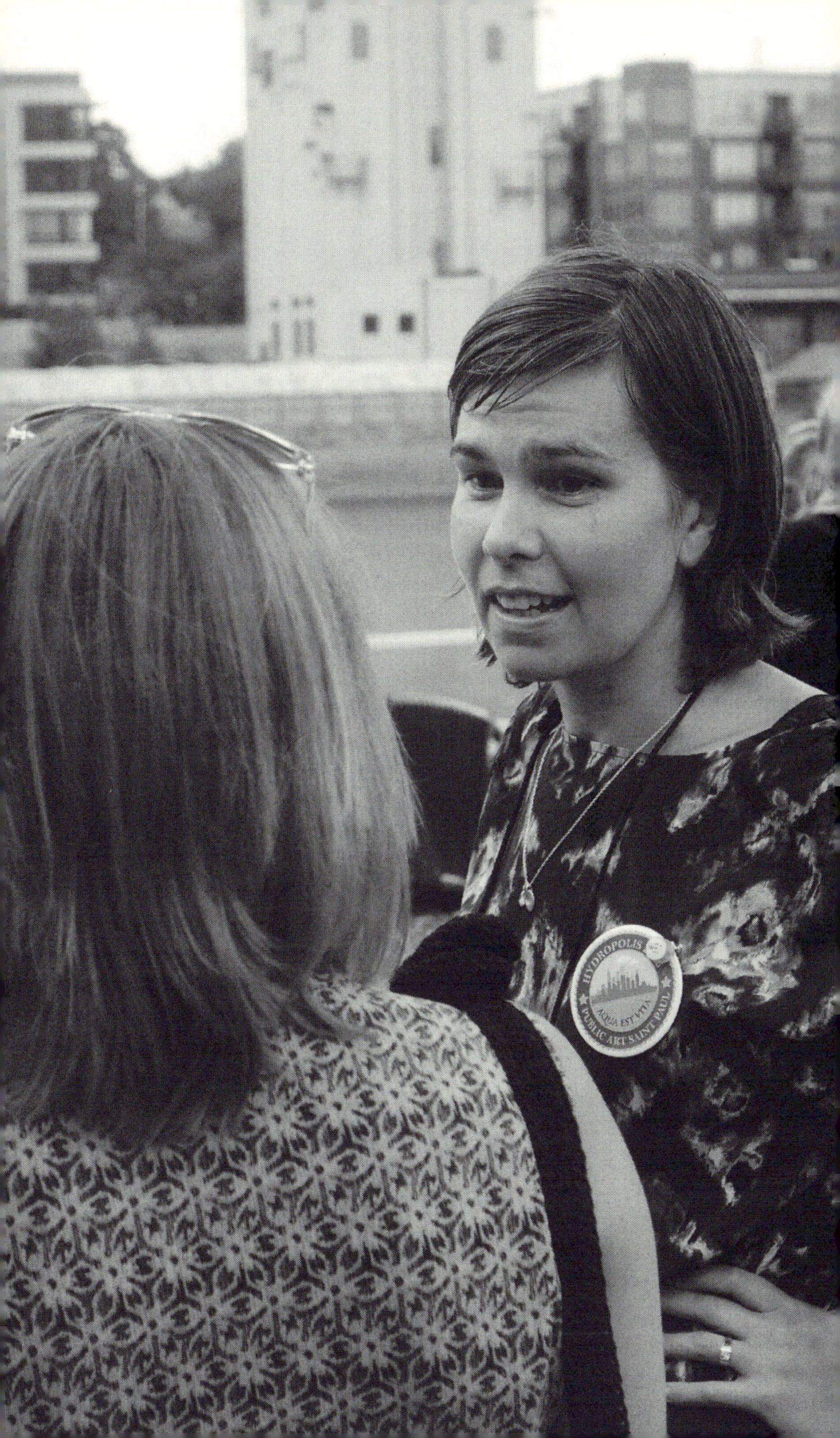
HYDROPOLIS
AQUA EST VITA
PUBLIC ART SAINT PAUL

Preceding pages: Collaboratory Science Fellow Kirsten Valentine Cadieux, artist Seitu Jones, Collaboratory Art Fellow Amanda Lovelee, and artist Ashley Hanson talking aboard a riverboat chartered by Public Art Saint Paul in conjunction with the LID Symposium in 2013. Photo: Sean Smuda.

Reflection: Arising From the River

Christine Podas-Larson

Founder of Public Art Saint Paul Christine Podas-Larson reflects on the many origins of City Art Collaboratory, including the ongoing efforts of Public Art Saint Paul to bridge conversations between public artists, environmental scientists, and our river city.

City Art Collaboratory evolved from Public Art Saint Paul's abiding love of the Mississippi River, the source of our city and the inspiration for our founding nearly 30 years ago. As far back as 1993, artist James Carpenter opened for us the possibility of light in the river valley with an extraordinary V-Mast cable-stayed bridge design that sprang from his deep understanding of the science of light, his lifetime on and observing rivers and seas, and his practice of respectful collaboration with scientists and engineers. The idea of the V-mast changed everything for us: it offered an illumined grace of vision and gratitude for our river; it slow cooked in our minds over time. Public Art Saint Paul began its environmental program in earnest 12 years ago. Critical to this initiative was our partnership with Eureka Recycling to plan and produce a zero waste international stone carving symposium and to realize City Artist Marcus Young's Wishes for the Sky as an aspirational annual earth day event on the Mississippi. The vital dialogue between artists and Eureka's environmental scientists and advocates presented us with new awareness and impetus for action.

City Art Collaboratory evolved from our presence in the midst of working artists in Lowertown's Jax Building. For so many of those artists, the river was essential to their experience of our city; for so many, awareness of environmental conditions challenged artistic practice with its bold dare to pursue sustainable artmaking for the sake of that beloved river. Artists Cliff Garten and Christine Baeumler spoke to us of the expressive potential of our watershed that would reveal water systems and change human behavior. Sustainable Artmaking Fellow Olive Bieringa collaborated with Marcus to bring artists and STEM professionals together to wander and wonder together along the Mississippi. Artist Shanai Matteson has been the inspired pilot of the City Art Collaboratory adventure since its inception.

Finally, City Art Collaboratory evolved from the idea at the core of Public Art Saint Paul's work from the beginning: that artists in the public sphere flourish when they collaborate with professionals across disciplines. City Art Collaboratory's artists and scientists allow themselves to be stimulated by shared curiosity and experience and by newfound understanding. In their own ways they go forth in creative scientific and artistic action to have rippling impacts throughout their networks and communities. In the end, their wanderings and wonderings and work are all about this river from which our city arose–this river that de Tocqueville called "the most magnificent habitation ever prepared by god for man."

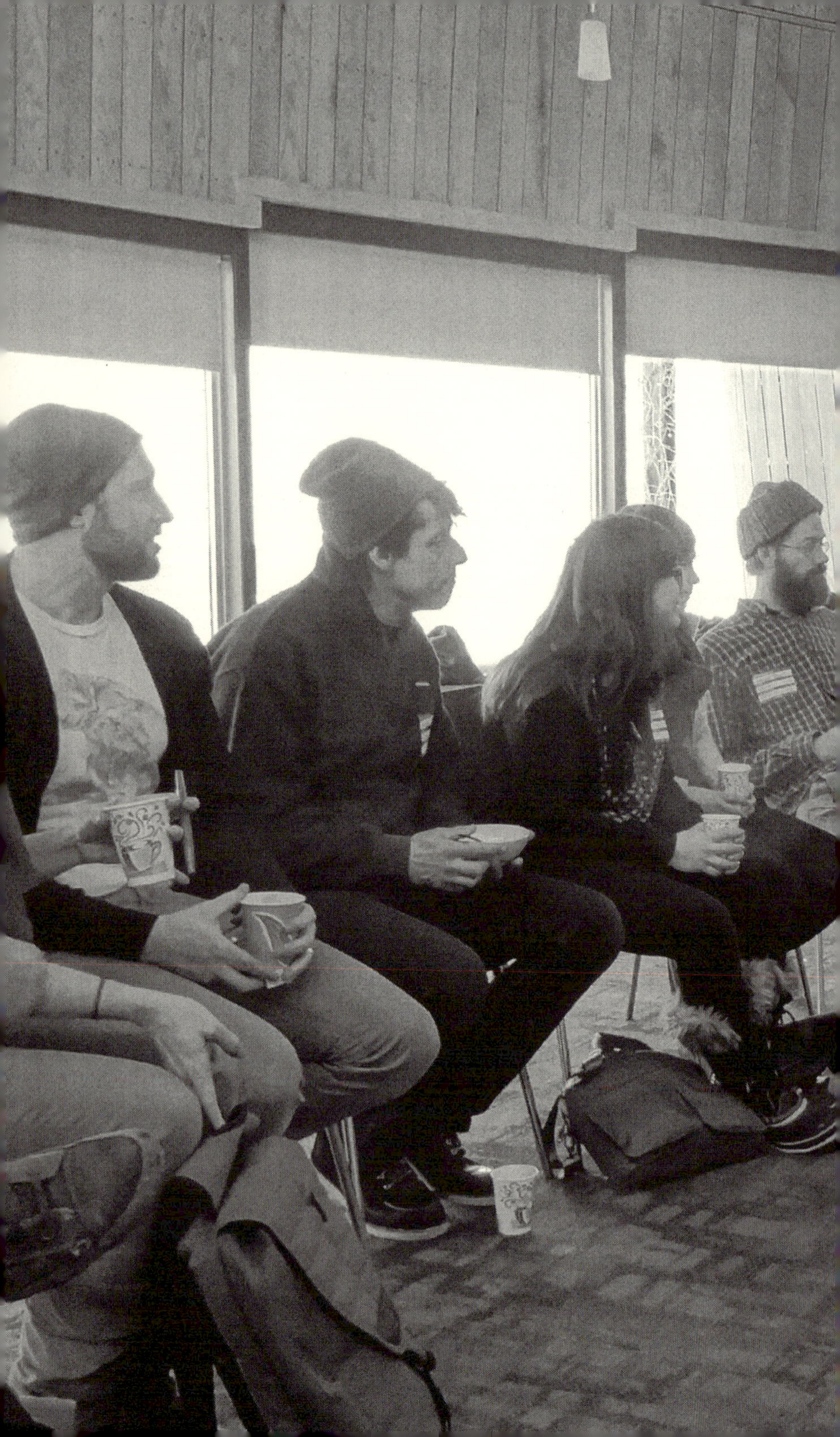

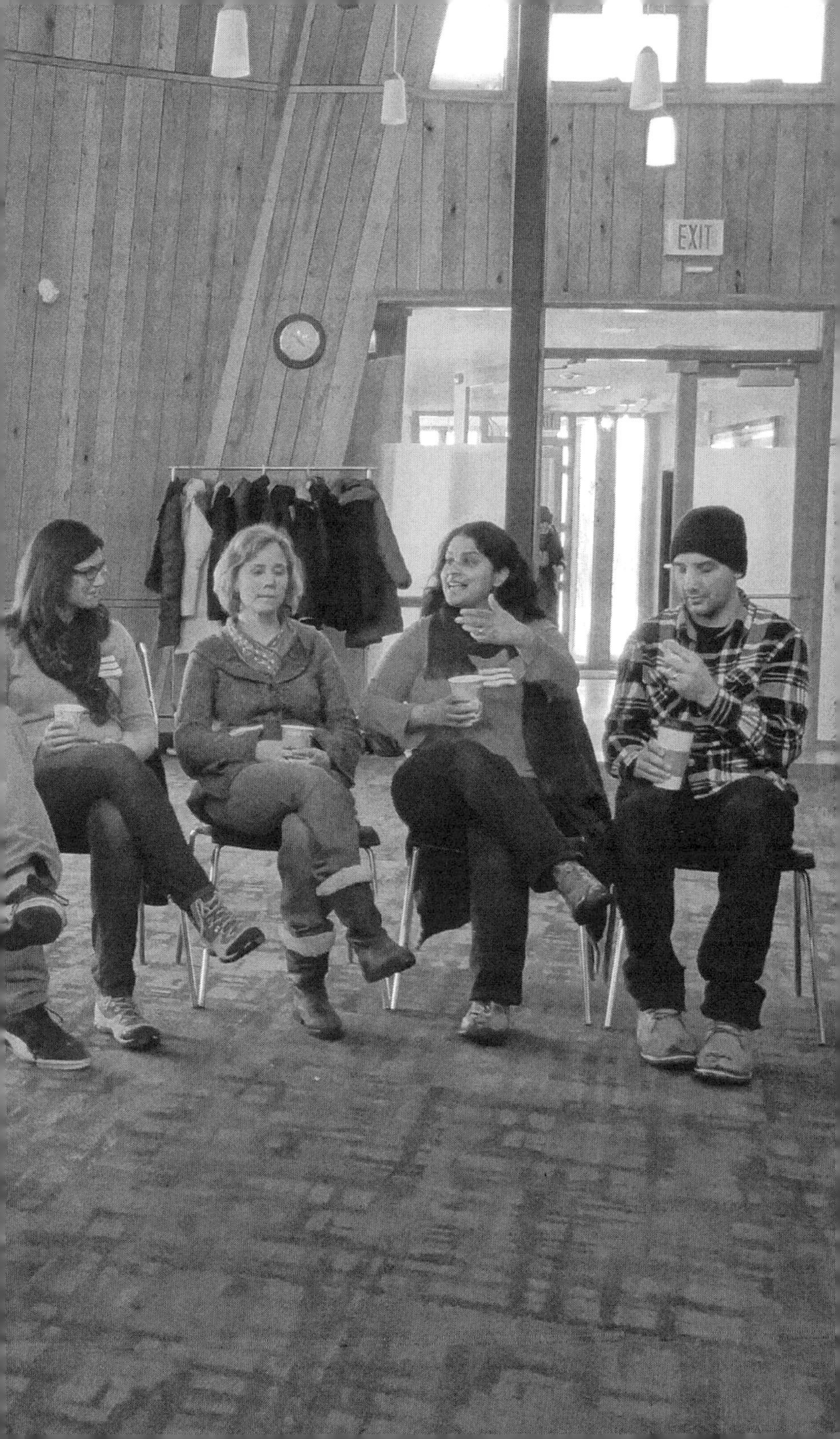
EXIT

Preceding pages: Collaboratory Art and Science Fellows meet for a retreat at Schaar's Bluff Gathering Center. Photo: Andrea Steudel.

Cohort & Contributor Biographies

Collaboratory Art Fellows

Molly Balcom Raleigh makes participatory performance and installation through media, including food, singing, found objects, and vernacular texts. Through a widely-varied practice, she seeks to surface the places within large social systems where individual actors, working solo or collaboratively, can effect radical change and regain agency. She has presented her work at many public sites and galleries, including Northern Spark, The Soap Factory, the Walker Art Center, The Nash Gallery, and The Wormfarm Institute. She is a 2014-2015 *Art(ists) on the Verge* fellow and co-founder of Prairie Fire Lady Choir. Balcom Raleigh studied playwriting and directing at the University of Minnesota and holds an Individually-designed Bachelor of Arts in Food Studies and Performance Studies from Goddard College. A resident of the Twin Cities for 20 years, she currently lives and works in Finland.

Aaron Dysart's work uses wry humor to push viewers to engage in the permeable division between the manufactured and non-manufactured. His recent environmental interventions and sculptural objects push ideas of propriety, gift giving, and reciprocity, while showcasing his love of material and it's ability to carry content. Aaron received Public Art Saint Paul's Fellowship for Sustainable Art-Making, took part in the Hunting and Gathering Walks program, and is now one of two City Artists with the City of Saint Paul through Public Art Saint Paul. He has shown nationally and received awards from Franconia Sculpture Park, Forecast Public Art, The Knight Foundation, and the Minnesota State Arts Board. He lives and works in Northeast Minneapolis, where he strives to understand what it means to be a small part of an all-engulfing system.

As an artist and activist, **Katie Hargrave** use community-oriented research to investigate the creation of mythologies, the historical record, and changes in the landscape. She makes her research visible by creating artwork that allows various perspectives and stories to come into contact. The outcome of her research varies in every project, including performance, social-practice events, printed matter, installation, sound, and video. Her projects ask the viewer to unwind a puzzle and realize her place as an actor in complicated national histories and present-day systems. Crucial to her practice, she asks the viewer to step away from simply observing to write her part of the story. Katie has developed an interest in seeing complex environmental systems in urban and rural environments.

Amanda Lovelee is a visual artist based in Minnesota currently working as a City Artist for Public Art Saint Paul housed within the City of Saint Paul. Lovelee acts as translator between the City's ideas and its residents, with the goal of building a city everyone wants to live in. As City Artist, Lovelee has focused on civic engagement through both her projects *Pop Up Meeting*, a city popsicle truck, and *Urban Flower Field*, a vacant lot turned gathering space/live science project. She is interested in how people connect and the spaces in which they do so within contemporary society. Her personal projects have explored myriad topics: the lives of beekeepers and ice fishermen, strangers' love stories, and the sociology of square dancing. She has an MFA in Visual Studies from the Minneapolis College of Art and Design and BFA in Photography from University of Hartford.

Aaron Marx is an artist, architectural designer, and educator working at the intersection of computer and video technology, architecture, and public space. Trained as an architect, with a background in mathematics and literature, Aaron teaches design in the School of Architecture at the University of Minnesota, creates designs for residential and commercial architecture, makes art in the public realm, and strives to investigate social issues through artistic practice. Aaron is an active member of Minneapolis Art on Wheels (MAW), Makesh!t collective, and City Art Collaboratory. Through the Collaboratory, he further explored the connections between scientific and artistic thinking by developing collaborations with other artists and scientists interested in technology, environment, and public space.

Peter Morales has a background in biology, foreign languages, and visual arts. He applies his multiple interests to sculpture. Morales's work is sited in various public locations throughout Minneapolis and Saint Paul. He currently lives and works in Saint Paul, Minnesota.

Janaki Ranpura makes art installations meant to be used and touched. The physicality guides people to connect with or intentionally disconnect from their surroundings. Stimulating tactile sensation and proprioception leads the participants in her projects to be surprised by how they inhabit their bodies and how their bodies together inhabit space. Her work includes a marionette show about abuses of power performed for an audience squashed into a renovated garage, a nighttime hide-and-seek game of costumed sperm searching for a 6' tall roving egg, a multi-room installation about uses of democracy in connection with the Greek referendum on staying in the EU, and a collaborative book-making project for the Nobel Foundation. She works with museums, theaters, galleries, universities, and local governments to stimulate visceral connections to social and policy issues. She currently helps run a sustainable technologies and design studio in Barcelona, Spain.

Emily Stover is a designer and public artist working in alternative landscapes and temporary architecture. She is interested in revealing the hidden social and physical phenomena that exist in the everyday environment, and in creating a public experience of the complex infrastructures of the city. Past projects include an interactive light installation on an energy utility, a temporary structure for making and sharing filled dumplings, and an inflatable public space for transportation-related programming. In addition to her practice, she is an instructor and researcher at the University of Minnesota College of Design, focusing on creative problem-solving and innovation in the public sector.

Asia Ward works between the realms of science, art, and education. Asia is a local artist who creates public art projects about the environment, water systems, and electrical power generation. Since 2005, Asia has been working for the Science Museum of Minnesota as Artist in Residence, Education Coordinator, and Program Specialist, and in 2012 she started working for the KidWind Project. Recently Asia became Co-Director of REcharge Labs, continuing her work developing products, writing curriculum, and training teachers about renewable energy education. Asia has participated in several artist residencies around the U.S. and has received grants from the Minnesota State Arts Board, Northern Lights.mn, Public Art Saint Paul, and the Knight Foundation.

Collaboratory Science Fellows

Kate Brauman studies interactions between water resources and land use, doing research motivated by people's need for both functional landscapes and safe, plentiful water supplies. Kate works at the University of Minnesota's Institute on the Environment (IonE), where she's the Lead Scientist for the Global Water Initiative. At IonE, Kate is evaluating how much benefit people get when using water and identifying where they could be getting more. In particular, Kate has focused on the global distribution of water use in agriculture and at the effects of land use on downstream water quantity and quality. For her Ph.D. research, Kate designed and ran a field study on the Big Island of Hawai'i to quantify the impact of land cover on water availability and explore whether water users could manage their water supply by paying upstream landowners.

Kirsten Valentine Cadieux explores social and spatial relationships related to landscape and food production. Using art and social science approaches, she builds publicly-engaged participatory research processes for people to learn about differing ways of understanding environments, and to practice performing and justifying environmental and food system interventions in collaborative ways. Her writing and teaching focus on how social and environmental practices build a basis for people to negotiate their aspirations for equitable, healthy, and interesting food systems and residential landscapes. She is currently working on the FoodShed, a collaborative online workshop and field guide to food movement efforts, and an accompanying podcast called *Eating Together*.

Jerald Dosch is a terrestrial ecologist with a particular interest in birds, non-native species, and the ecological challenges of the winter season. His research projects have included investigating the impacts of heavy metals (airborne mercury pollution and lead fishing tackle) on Common Loons in Minnesota; the importance of fruit-eating birds to rainforest regeneration in abandoned Costa Rican pastures; carbon storage in forests, and the role of non-native garlic mustard in Minnesota forests. Dosch teaches courses in ecology, outdoor environmental education, and environmental science.

Leili Fatehi is a legal and public policy expert with ten years of experience advancing research, facilitating dialogue, and catalyzing solutions at the nexus of science, technology, and society. Her specialties include law, science and technology; ethical, legal, and societal implications of emerging technologies; science and technology regulation and oversight; nanotechnology; genetics; robotics; and human enhancement. She is the owner and principal of Apparatus, a full-service public affairs firm focusing on complex issues at the nexus of social, natural, and built systems.

Adam Kokotovich is a postdoctoral research associate at the University of Minnesota's Minnesota Aquatic Invasive Species Research Center. He is an interdisciplinary social scientist who studies decision making concerning science, technology, and the environment. He is particularly interested in how the assumptions we hold concerning what constitutes knowledge, harm, and the desired state of the world influence decision making and the frameworks we use to inform decision making, such as risk assessment. In his current project he studies the management and risk assessment of invasive Asian carp. His dissertation research explored contestations of risk and risk assessment involving plant genetic engineering.

Matt Kumka is a landscape architect based in the Twin Cities. Since graduating with a Master's Degree in Landscape Architecture from the University of Minnesota, Matt has specialized the creation of green infrastructure, low-impact development, and restorative place-making at Barr Engineering, where he works closely with many local municipalities and watershed districts. With a specialty in water resource management he has worked to develop thoughtful landscapes along the continuum from cityscape through exurbia and back again. He is perpetually exploring the relationship between the natural elements of stone, soil, water, animals, and plants in order to spark humankind's intrinsic connection with our living world. Matt's goal with the Collaboratory is to help create lasting connections between those that organize and maintain our vital infrastructure and wild places, and the artists and thinkers that vivify and enlighten our world.

Daniel Nidzgorski fell in love with ecology out in the deep woods. He received a Ph.D. at the University of Minnesota, studying how urban trees affect the water quality of local lakes and streams. Daniel enjoys using urban ecology to bring science out into public view and right into people's neighborhoods, where he gets many questions about why he's digging holes in the city parks or collecting leaves from street gutters. He is also passionate about working to increase diversity and inclusion in science and engineering, including founding Minnesota Queer Science to help shatter the glass closet.

Bunmi Odumuye is a mechanical environmental engineer and the CEO of the Renewables Research and Policy Institute, an energy engineering research firm that promotes renewable energy systems and conservation measures. In addition, he is active in the Twin Cities food justice and aquaponics communities, facilitating an aquaponics lab at North High School in association with Project Sweetie Pie, and serving on the organizing committee for the Institute for Agriculture and Trade Policy's ongoing food + justice = democracy effort.

New Jersey native **Ilana Percher** currently resides in Minneapolis as she works towards a Ph.D. in experimental condensed matter physics. Her thesis work is a study of superconductivity in thin films. Ilana was a visual artist in a past life (and hopes her future will include more art-making, too). She has shown collaborative works as part of Version Festival, the Art Shanty Projects, and at Roots and Culture Gallery. Her exploration of what it means to be a member of both the science and art communities has been ongoing, and she has presented physics talks at Art Shanty Projects and the Northern Spark Festival. She considers ice cream to be her spirit food and urban berry picking her favorite summer activity.

Roopali Phadke is a broadly trained social scientist interested in how publics engage in environmental decision-making. She is a Professor of Environmental Studies at Macalester College. Her research and teaching are at the nexus of environmental policy, science and technology studies and participatory planning. Her projects have examined social activism against large dam development, the aesthetics of wind energy design, visualization and deliberation in climate change activism and the role of citizen photography in future cityscape design.

John Schade is an Ecologist who studies feedbacks between climate change and aquatic ecosystems, mostly in the Arctic in Alaska and eastern Siberia, but also in the forests of New Hampshire and agroecosystems in the midwestern U.S. He spent ten years as a professor of Biology and Environmental Studies at Saint Olaf College in Northfield, MN. Currently, he is a Program Director at the National Science Foundation and a Distinguished Visiting Scientist at the Woods Hole Research Center. In recent years, he has become deeply interested in developing collaborations between natural scientists, social scientists, writers, and artists to develop ways to work in interdisciplinary teams to solve problems, create curriculum, and communicate with the public.

Currently a Program Manager with the Freshwater Society, **Leslie Yetka** has worked for the past eighteen years in the areas of education, community engagement, and public planning with a focus on water resource management, sustainable landscape practices, and community resilience to climate change. She has worked with a variety of audiences, including private landowners, businesses, non-profit groups, professional staff, and community groups engaged in protection of our environment. Leslie has a bachelor's degree in biology from Macalester College, and a master's degree in horticulture with a minor in water resources science from the University of Minnesota, and is a lifelong resident of Minnesota where she enjoys the outdoors with her family and two poodles.

Collaboratory Organizers & Advisors

Olive Bieringa is a dance artist and somatic movement therapist. For the past eighteen years she has collaborated with Otto Ramstad as the BodyCartography

Project. The mission of BodyCartography is to engage with vital materiality of the body to build live performance, which facilitates a re-enchantment of embodiment, relationship, and presence. Over the past eighteen years they have made performances, installations, films, festivals and curatorial projects in public spaces, wilderness, museums and theaters which have been presented across the world supported by institutions such as the Foundation for Contemporary Arts, Bush Foundation, Walker Art Center, and Lyon Opera Ballet amongst others. Olive is a co-founder and curator of SEEDS Festival, an interdisciplinary arts and ecology festival, in Massachusetts. She inspired Public Art Saint Paul's Hunting and Gathering Walks and co-curated the City Art Collaboratory. She who holds an MFA in Performance and New Media from Long Island University and is a certified teacher of Body-Mind Centering®.

Shanai Matteson is an artist, writer, and arts organizer who develops, creates, and enables collaborative public art and design projects. She is interested in creativity and connection at the margins of established fields and practices, and believes in elevating everyday expertise, as well as creating more opportunities for artists to learn through collaborations with and in community. Shanai is one of the developers and Director of Public Art Saint Paul's City Art Collaboratory. Matteson is a 2011 alum of the Creative Community Leadership Institute at Intermedia Arts, and a 2013 Bush Fellow, and a 2014 River Stewards Emerging Leader.

City Art Collaboratory Advisor **Patrick Nunnally** coordinates the River Life program, which brings together people working on a sustainable, inclusive future for the Mississippi River and University of Minnesota researchers and teachers. Nunnally has worked on a wide assortment of Mississippi River programs since the 1990s, and has taught part time at the University since 2002. He is editor of *Open Rivers: Rethinking the Mississippi River*, an online journal that considers issues of water and equity from a variety of perspectives.

Andrea Steudel is a cultural events producer, project manager, and artist who specializes in creating mobile platforms for broadcasting technology-based media. She is a founding member of Minneapolis Art on Wheels and a producer of Northern Spark, an annual overnight festival of temporal public art. She has created work for Forecast Public Art and the Charlotte Street Foundation. She is holds a B.A. in Experimental and Media Arts (University of Minnesota), and a Sustainability Studies minor. Andrea assisted with organizing and administering the City Art Collaboratory.

Marcus Young/楊墨 is a behavioral and social practice artist making work for the concert stage, museums, and the public realm. His work expands the repertoire of human behavior and the expressivity of the city. As a Saint Paul City Artist (2006 - 2015), Marcus redefined the role of artists working within city government. His project *Everyday Poems for City Sidewalk* transformed the city's sidewalk maintenance program into a publishing entity for poetry. He also co-led the development of City Art Collaboratory as a public art curator, artistic advisor, and participant. Interested in DIY spirituality, he created *Don't You Feel It Too?*—a participatory street dance practice of social healing, civic belonging, and inner-life liberation. He has led more than 200 sessions of this dance practice for art festivals, schools and colleges, youth experiencing homelessness, professional dance groups, and the general public. He is ongoing Collaborating Director with Ananya Dance Theater, and a recipient of awards from the McKnight Foundation, Bush Foundation, Jerome Foundation, Franklin Furnace, and Forecast Public Art.

Other Contributors

Melissa Constantine is a writer, editor, artist, and designer based in New York City. She is a member of the artist collective Complimenta, a group that facilitates events, residencies, and publications as forms of inquiry around social, economic, ecological, and creative practices. She is a Ph.D. candidate in Anthropology at Columbia University, and previously received a B.A. in Art and Philosophy from Hampshire College and M. Arch from Cornell University. Her current research engages art collectives and theories of collectivity to consider the affective and material force of nascent, relational, and intimate political subjectivities.

Adam Kay is Director of Social Innovation and Associate Professor of Biology at the University of Saint Thomas. His research is on "Stewardship Science," which combines environmental science experiments with community service. Most of this work involves urban agriculture and urban ecology. Adam is the co-founder and director of BrightSide, a sustainable business for delivering fresh produce to urban corner stores, and the UST Stewardship Garden, a community garden that integrates biological research with a food donation program.

Colin Kloecker is an artist, designer, and filmmaker working to make cities and communities more resilient, playful, and supportive by amplifying and nourishing everyday knowledge and creativity where it already exists. He founded the nationally recognized and exhibited public art and design studio, Works Progress. Kloecker is a 2011 Fellow in the Creative Community Leadership Institute and a 2014 River Stewards Emerging Leader. As an artist he has received awards and commissions from the Walker Art Center, Forecast Public Art, the Minnesota State Arts Board, the Saint Paul Riverfront Corporation, and the City of Saint Paul's Public Art Program, and has exhibited work at Blue Star Contemporary (San Antonio, TX), Elsewhere (Greensboro, NC), and Crystal Bridges Museum (Bentonville, AK). In 2015, he co-directed Hand-in-Glove 2015, a national conference for socially engaged arts organizers.

In 1987 **Christine Podas-Larson** founded Public Art Saint Paul and served as its President/CEO and Executive Producer for over 28 years. She forged and sustained a nationally unique public/private partnership with the City of Saint Paul and was the chief author of Saint Paul's cutting edge public art ordinance. She led Public Art Saint Paul's public information efforts to deeply engage the community and reach worldwide media audiences. Her interest in big ideas of artists working in the public sphere brought forth major projects with lasting impact on the shape and future of the city and the everyday life of its citizens. Her work has been lauded by Mayor Chris Coleman, Senator Al Franken, U.S. Rep Betty McCollum, former Vice President Walter Mondale, MN Rep Richard Cohen, and many others. Her arts leadership has been the focus of numerous media features. Christine is deeply involved in the community and has a long volunteer history that includes service on numerous non-profit boards of directors.

Christina Schmid thinks with art and writes as critical practice. Her essays and reviews have been published both online and in print, in anthologies, journals, artist books, exhibition catalogs, and digital platforms. She works at the University of Minnesota's Department of Art in Minneapolis, where she gets to talk about art for a living. Her teaching focuses on contemporary art, mixed with critical theories, framed by cross-cultural currents, and driven by deep curiosity. She is passionate about getting lost in the woods, running with dogs, and drinking sweet rose tea.

Colleen Sheehy became Executive Director of Public Art Saint Paul in 2015. She holds an MA and PhD in American Studies from the University of Minnesota. Her studies brought together the disciplines of art history, folklore, cultural geography, urban studies, and popular culture. Her professional work has been concerned with ways in which art and artists play vital roles in civic life. From 2008 to 2015, she served as Director and CEO of Plains Art Museum, where she led museum-generated public art and community engagement projects and played a leading role in conversations on public art and social practice. She was a member of the Fargo Public Art Task Force, which helped to pass an ordinance to establish an Arts Commission, and she served as an appointed member on the Commission. Prior to that, she was Director of Education at Weisman Art Museum, University of Minnesota for 15 years, where she forged collaborative models of museum education that linked cross-disciplinary learning with campus and community audiences.

Dan S. Wang/王念华 works in the global midwest. His critical writings have been published internationally in journals, exhibition catalogues, and book collections. He has lectured in many places, including the Salzburger Kunstverein, Art Institute of Chicago, Rockbund Art Museum (Shanghai), *documenta 12* (Kassel, Germany), the Central Academy of Fine Arts (Beijing), and the third Creative Time Summit (New York). He frequently produces traditional printmedia for circulation through art networks and in functional activist situations. Dan lives in Madison, Wisconsin, where he works with Madison Mutual Drift, a group that writes about racial politics, city space, and movement strategy.

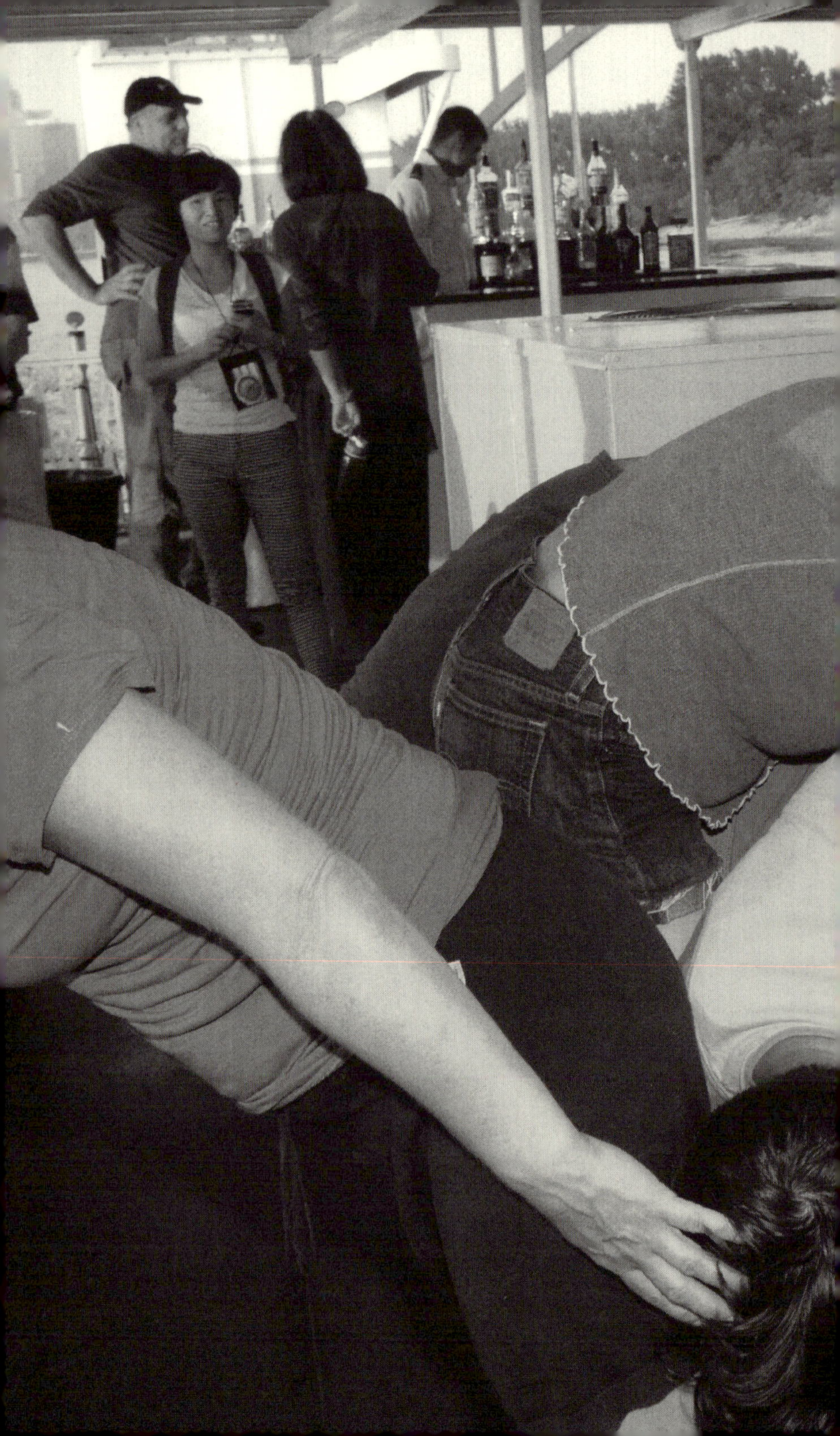

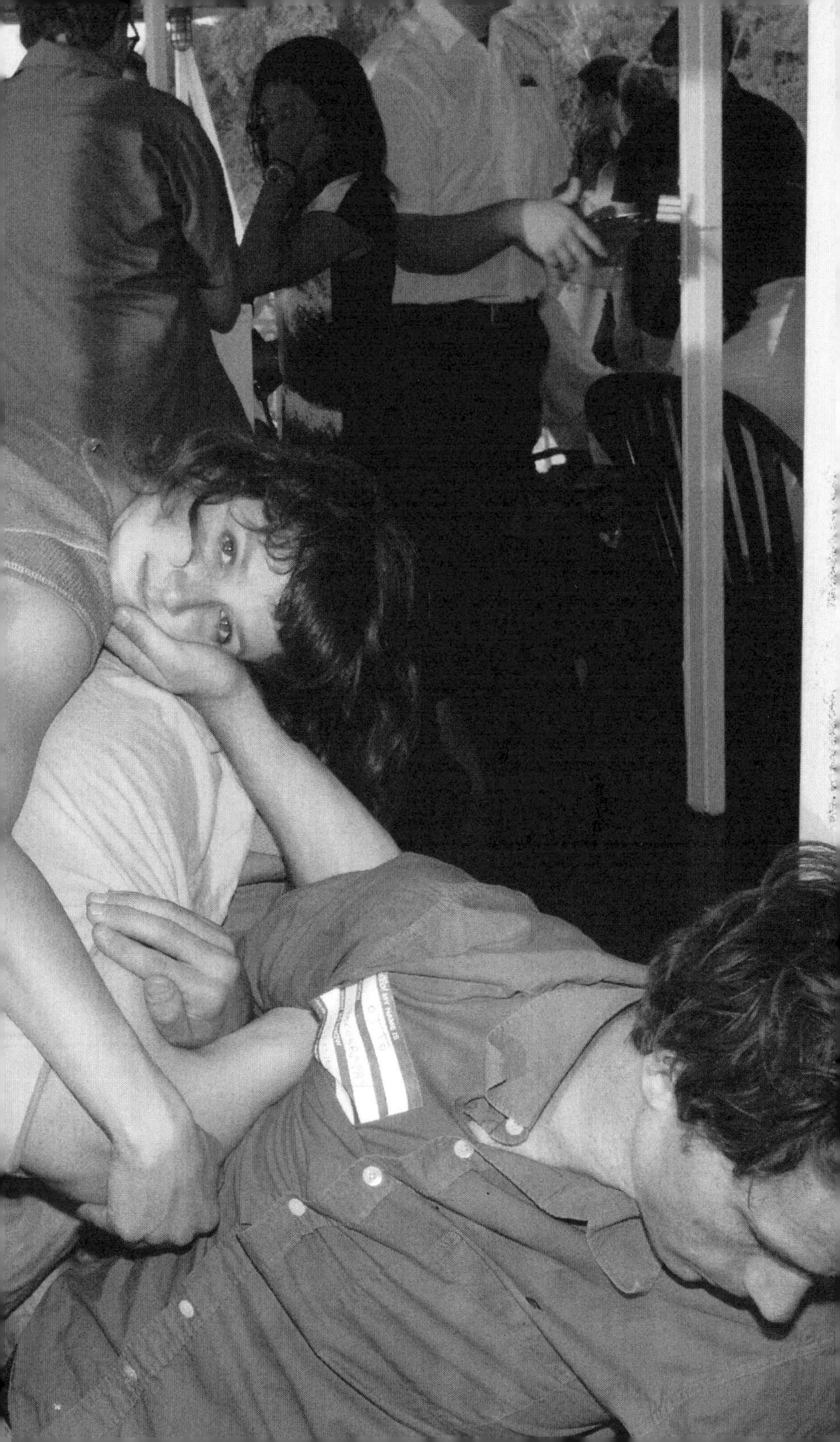

Preceding pages: The BodyCartography Project performing aboard a riverboat chartered by Public Art Saint Paul in conjunction with the LID Symposium in 2013. Photo: Sean Smuda.

Acknowledgments

Launched in 2012 by Public Art Saint Paul with support from the National Endowment for the Arts, City Art Collaboratory has become a nationally unique and locally vital space for artistic inquiry, direct experience of our urban environment, critical reflection on ecologies of place, and the production of environmentally focused public art and engagement.

The Collaboratory grew out of Public Art Saint Paul's environmental art initiatives of 2008-10, including the Sustainable Artmaking Fellowship and Hunting and Gathering Walks. Environmental artist Shanai Matteson (Works Progress Studio) became director of the Collaboratory; Olive Bieringa (BodyCartography Project) and Marcus Young (Public Art Saint Paul City Artist) served as curators, and were also the first to develop and to help launch this program.

The initial City Art Collaboratory cohort included eight public artists and eight scientists, and expanded in 2013 with the addition of several new members. Thank you to those participants, including artists Molly Balcom Raleigh, Aaron Dysart, Katie Hargrave, Amanda Lovelee, Aaron Marx, Peter Morales, Janaki Ranpura, Andrea Steudel, Emily Stover, and Asia

Ward; and Science Fellows Kate Brauman, Kirsten Valentine Cadieux, Adam Kokotovich, Matt Kumka, Bunmi Odumuye, Roopali Phadke, Leslie Yetka, Daniel Nidzgorski, Jerald Dosch, John Schade, and Ilana Percher.

Thanks also to those who advised this program, including Patrick Nunnally at the University of Minnesota's River Life program, artist Christine Baeumler, and the many others who took part in Collaboratory field trips and workshops.

Thank you to Christine Podas-Larson, Founder and former President of Public Art Saint Paul, for her leadership and willingness to support the passion and vision of public artists. Thanks also to Public Art Saint Paul staff (Colleen Sheehy, Nic Cherry, Mary Johnson, Amanda Lovelee, Aaron Dysart, as well as past staff Eric Asboe, Amy Krautbauer, and Ashley Hansen) for facilitating this work, and to Public Art Saint Paul's Board of Directors (Edward F. Fox, Joan Palm, Heather Peterson, Peter Kramer, Ahmed Tharwat. Tom Eggum, Pierce Canser, Emmy Treichel, Susan David Price, Nick Fox, Andrea Stimmel, Bernie Bullert, Toni Carter, Rich Lallier, Bob Bierscheid, Bob Muschewske, Yamy Vang, Marilyn Porter, Sarah Schultz, and Mayor Chris Coleman).

And finally, thank you to our funders at the National Endowment for the Arts—ArtWorks, District Energy, The Saint Paul Foundation, F.R. Bigelow Foundation,

Mardag Foundation, The John and Ruth Huss Fund of The Saint Paul Foundation, and The voters of Minnesota through a Minnesota State Arts Board Operating Support grant, from a legislative appropriation of the arts and cultural heritage fund.